CHILD | DATA | CITIZEN

CHILD | DATA | CITIZEN

HOW TECH COMPANIES ARE PROFILING US FROM BEFORE BIRTH

VERONICA BARASSI

The MIT Press
Cambridge, Massachusetts
London, England

This book was set in Bembo Book MT Pro by Westchester Publishing Services. Printed and bound in the United States of America.

Library of Congress Cataloging-in-Publication Data

Names: Barassi, Veronica, author.
Title: Child data citizen : how tech companies are profiling us from before birth / Veronica Barassi.
Description: Cambridge, Massachusetts : The MIT Press, [2020] | Includes bibliographical references and index.
Identifiers: LCCN 2020015011 | ISBN 9780262044714 (paperback)
Subjects: LCSH: Children--Research. | Data mining. | Consumer profiling. | Citizenship.
Classification: LCC HQ767.85 .B37 2020 | DDC 305.23072--dc23
LC record available at https://lccn.loc.gov/2020015011

10 9 8 7 6 5 4 3 2 1

For my daughters

CONTENTS

ACKNOWLEDGMENTS

I wrote this book for my daughters, my family, and all the parents and families I met in this life-changing and personal research journey. I owe them everything. Their stories, their thoughts, and their everyday experiences have shaped my understanding of the injustice of surveillance capitalism. They made this book possible; they inspired me, questioned me, surprised me, and reminded me of why we need publicly engaged ethnographic research. I doubt that I will ever be able to convey how much I owe them.

My immense gratitude goes also to my colleagues and friends who have supported me throughout the way. I am thankful for their work, amazing minds, advice, collegiality, critical thinking, and passion for research. Without their support, I would have never been able to finish this project and book. Again there are no real words to express my gratitude.

I would like to thank in particular Lina Dencik, Mila Steele, and Melissa Sevasti-Nolas for being such amazing friends, believing in this book, and making it better with their observations, thoughts, and ideas. I also would like to thank Ben Williamson for his generosity and comments. Without their input this book would have been very different.

The Child | Data | Citizen project would have not been possible without the academic and emotional support of different scholars and friends. I personally would like to thank, Natalie Fenton, Leah Lievrouw, Sonia Livingstone, Nick Couldry, Annette Markham, and Greg Elmer for being there for me during these life-changing times. They have been a constant source of motivation and support. I am deeply indebted to them.

In the last three years different colleagues have inspired me with their work and our discussions. A special thanks goes also to Adrienne Russell, Emiliano Treré, Alice Mattoni, Anastasia Kavada, Stefania Milan, Thomas Poell, Mirca Madianou, Gholam Khiabany, Des Freedman, Milly Williamson, Richard MacDonald, Angela MacRobbie, Clea Bourne, Aeron Davis, Lisa Blackman, Joanna Redden, Arne Hintz, Dan McQuillan, Elefetheria Lekakis,

Anne Kaun, Giovanna Mascheroni, Tama Leaver, Tom Boellstorff, Mimi Ito, and Sarah Pink.

I would also like to thank the Department of Media, Communications and Cultural Studies at Goldsmiths and the British Academy for having believed in this project and for giving me their institutional support. I am also profoundly grateful to my friend Craig Barrett for helping me with the design and running of the Child | Data | Citizen website, which has been an invaluable resource for this project.

All my gratitude goes to my family. It goes to my parents, Patrizia and Gianni, for their unconditional love, support, and for being the best examples I have in life. It also goes to my sister Brada—who is always there for me, listens, and encourages me all the time—and to my niece and Ale. Finally, I am really not sure where I can find the right words to thank Paul, my husband, who has read the book from cover to cover, commented, pushed me to achieve more, and always supported me no matter what. In the craziness of our cross-continental life, he has always been there together with my daughters, filling my life with joy and excitement. Thank you.

INTRODUCTION

> I feel like data is being collected all over and misinterpreted. I feel like they have so much faith in what their algorithms tell them, that you are no longer yourself, but you are just what the algorithm says you are. And it's all a guess right? . . . They sell you products or they tell you which candidate to vote on the basis of that data. But I don't think that data is accurate. . . . It's kind of the labeling theory. In the sense that you may start to believe who you are as an individual on the basis of what they say you are. . . . Everything is becoming more integrated. More and more. But you can't get away from the technology, because it's cheaper and easier; but at the same time you are signing off your personal data.
> —Carlos, west Los Angeles, graphic designer, two children aged 5 and 7 years

Carlos[1] is a designer; he grew up in central California from a family of Mexican-American farmers and moved to Los Angeles in his early twenties to start his career. When I interviewed him for the Child | Data | Citizen project, Carlos told me of his concern about the ways in which he and his children were being profiled because of their data. In the interview, I asked him to describe what it felt like when he thought about online privacy, especially in relation to his children. As I listened to what he had to say, I realized he couldn't have explained it better. That was what data profiling felt like to me too, the feeling that your data is constantly collected from you and that you have no choice. Either you sign off your personal information or you are cut out from an array of services and relationships. There is also that sense that you can't get away from technology. Everyday life is defined by that untold feeling that it is impossible to escape technologies that monitor and turn us and our children into data.

Examples are countless. Wherever we turn, we encounter a plurality of technologies that gather, archive, and process our personal data and the personal data of our children. We can think about the facial recognition software used by an airport security guard on our 10-month-old daughter or the smart basinet that our in-laws bought for their first grandson. We can also think about all the times that we signed off our personal data and the data of our children without reading the terms and conditions, which are often—and even after the introduction of the General Data Protection Regulation (GDPR) in Europe—too difficult, too long, and too vague to understand. Finally, there is that feeling of unease and worry as we start to realize that our personal data is constantly used to profile, sort, or turn us into data subjects.

Like Carlos, I and different parents I met during the project were starting to deal with these questions. Why is our family life being turned into data? How is this data being used? How are we being profiled? Are these algorithmic predictions accurate, or—like Carlos mentioned—are they just a random guess based on imprecise data traces? How will these algorithmic predictions impact our lives and the lives of our children?

We really don't have an answer to all these questions. Family life is immersed in complex and new data environments, which keep changing and evolving, especially with new developments in artificial intelligence technologies. The speed of technological change and the rapid transformation in business models makes it very difficult for us to have a clear picture of what happens to all the data that is collected from our digital interactions. In addition to this, what makes things more complex is that the business model of companies is dependent on the *secrecy of algorithms* (Pasquale 2016) and so we do not know how we are being profiled, sorted, and classified on the basis of our data traces.

Although it is clear that we are being constructed as data subjects and these constructions are impacting our lives in a variety of ways, it is also clear that this process of algorithmic definition and construction escapes our control (Cheney-Lippold 2017). Like Carlos, we do not know how we or our children are being constructed as data subjects. However, we know for sure that questions about algorithmic bias, systemic discrimination, and data harm are coming to the fore as more and more companies and governments use predictive analytics and AI systems to profile individuals and make data-driven decisions about them. Researchers, journalists, not-for-profit organizations, businesses, lawmakers, and individual parents like Carlos are all asking these questions.

This book engages with these questions by exploring current debates on algorithmic bias, AI ethics, and data rights. Its aim is to advance a new, distinct understanding that explains our lived experiences as parents and family members and situates this experience within a broader critical analysis of surveillance capitalism. My hope is to empower parents to seek legal change. I believe that the real problem that we face today as a society is that we are trusting technologies to profile humans, but as I will show in this book and as Carlos rightly mentioned, these technologies will always be inaccurate and biased when it comes to profiling humans. The human error in algorithmic profiling may be reduced but it can never be eradicated. Companies cannot fix the problem with a technical solution or by appointing a new AI ethics board, and individuals can't really protect themselves or their families. This is a systemic problem, and the only solution to this is a political solution. Governments must step up and recognize that our data rights are tightly interconnected with our civil rights; as parents we need to start working together as collectives, organizations, and institutions to demand a political change.

A PERSONAL RESEARCH JOURNEY

The idea for this project and book came to me in 2015. I am an anthropologist and mother of two young girls, and I suddenly realized that there are vast—almost unimaginable—amounts of data traces that are being produced and collected about children. This book is based on the findings of the Child | Data | Citizen project, a three-year-long research project that enabled me to investigate the multiple ways in which children's data traces are being produced, shared, and used.

I started designing the project when, in 2015, I had my first daughter (P). During the first year of my daughter's life, I was writing *Activism on the Web: Everyday Struggles against Digital Capitalism* (Barassi 2015), which was concerned with the relationship between social media and political identity and explored key questions about online privacy, digital surveillance, and big data. As you can imagine, writing a book, while being on maternity leave entails a lot of juggling. Between a feed, a nap, and the writing up of a chapter, I was introduced to the extraordinary world of new parents. We would meet regularly and enjoy different parent/child activities; we would organize picnics and eat a lot of cake. Meeting up with other new parents was not only life-saving because I could share the ups and downs of early motherhood, but it was also life-changing because it enabled me to see things from a different perspective.

As days went by, and my maternity leave was coming to an end (as well as my book), I started to realize how much personal data of children is produced from the moment they are conceived. Not only my fellow peers (and myself included) were recording important medical data on mobile apps, but we were extensively sharing photos of our children online through public and private social media platforms. As I was writing about online privacy, surveillance, and data cultures in the context of activism, I realized that we had very little understanding of what was going on within family life.

Everywhere I turned data were produced and collected about children, and at times this was very sensitive information that could have an impact on their future life. I remember being particularly interested in all those parents that were sharing images on social media of children at political protests. I remember my concern as I realized that the photos of children with banners criticizing the current government, which were posted on social media, could easily lead to political profiling from a very young age (Barassi 2017b). Hence I started wondering: How were children's data traces produced? How were parents negotiating with online privacy, data mining, and digital profiling? What type of data were companies collecting? Were companies profiling children from before birth?

It was to address these questions that I designed the Child | Data | Citizen project, and I started to work on children's data traces. After staring my project, I became pregnant with my second daughter (A) and Google knew I was pregnant before my family did! At that time, I found myself immersed in a new data environment in which children's data traces were collected each day by new agents and technologies (let's think not only about home hubs and artificial intelligence toys, but also new developments in facial recognition that have happened between 2015 and 2017). This was also a time of great change in my life for one main reason: my husband was relocated for work to Los Angeles. Hence we started living between London and Los Angeles dealing with two homes, two health and education systems, and of course two very different data environments.

As I was living between London and LA, I came to the conclusion that if I wanted to understand the multiple and complex ways in which datafication was transforming family life, I had much to gain if I considered the relationship between the UK and US contexts. Much of the digital technologies that families in the UK use today are produced within the US. Therefore, a cross-cultural analysis between UK and US contexts could enable me to shed light not only on how big data and digital surveillance were being negotiated in different social contexts, but also on the fact (or so I thought) that the use of the same digital technologies is often defined by completely different cultural understandings of online privacy.

So when I started the project, P, my first daughter, was 2 years old and I was pregnant with A. I hold a PhD in social anthropology, and I have always been a firm believer in the merits of the ethnographic method. In anthropology, the origin of the ethnographic method can be found in the works of Bronisław Malinowski (who conducted fieldwork in the Trobriand Islands and taught in England) and Franz Boas (who conducted fieldwork in Baffin Island and taught in the United States) who promoted it. Through the ethnographic method, the idea is for the anthropologist to actually live in a specific society and/or cultural setting for a considerable period of time, simultaneously participating in and observing the social and cultural life while gathering important information on the economy, politics, and culture. The basic idea of the method is to enable researchers to learn through direct experience, because the ethnographer finds him/herself immersed in a social reality and tries to make sense of it through the self.

Hence, armed with pen and paper, I started documenting how it felt to live in a world in which I, my family, and children were being datafied, but also as a parent, I had very little control over how the data of my children were produced and shared. Hence, I became a participant observer in my own life and I started taking notes. I started writing fieldnotes about how it felt when random parents at the park took photos of my child (simply because she was sitting next to theirs and without asking my consent) or how distressing it was to have to sign terms and conditions, even if I knew that my consent was somehow coerced and certainly not informed. I started to observe what other parents were doing in parks, school meetings, children's parties, playdates, and all the other events that were key to my life as a parent.[2] This auto-ethnographic research, therefore, enabled me to start tackling the lived experience of the datafication of children from a parent's perspective.

Although there is certainly a sensory dimension to the ways in which we can understand the experience of datafication, and hence auto-ethnographic research is particularly interesting, I also wanted to investigate how different parents made sense of the experience. Therefore, I worked with families in London and Los Angeles, with children between 0 and 13 years of age whose personal information online is ruled by the Children's Online Privacy Protection Act (1998). I carried out fifty semistructured interviews and eight months of digital ethnography of eight families "sharenting" practices on their social media accounts.

In both cities, I worked with parents who came from a variety of cultural, ethnic, and national backgrounds. The parents were extraordinarily diverse not only in terms of ethnicity (e.g., Asian, Latinos, Indian, black, indigenous, multiracial, white) but also in terms of cultural and national heritage (e.g., Afghani, Mexican, Brazilian, Indian, German, Italian, Hungarian, Icelandic, Zimbabwe, Scottish). I also made a genuine attempt to seek parents from different classes by interviewing parents working on low-income jobs (such as nannies, cleaners, buskers, or administrators) as well as parents working in high-income jobs (such as lawyers, film producers, journalists, marketers). I also came across a plurality of family situations that challenged the heteronormativity of the nuclear family, and I interviewed gay parents, parents who were divorced and had to juggle with a complex living arrangement, and single mothers who chose to adopt a child.

Although I was interested in understanding the human experience of the datafication of childhood, I was also determined to try and make sense of the ways in which children's data was being collected, used, and profiled by companies. It is for this reason that I decided to complement my findings with a "platform analysis" of four social media platforms, ten health-tracking apps (baby apps and pregnancy apps), four home hubs, and four educational platforms. The analysis consisted in the study of the promotional cultures, business models, and data policies of the different platforms. It also consisted in researching and following the different news that involved patent applications as well as privacy scandals. The platform analysis was largely qualitative and ethnographically informed in the sense that I found myself analyzing these platforms simultaneously as a researcher and as a concerned parent who wanted to find out how my children's data (and my own) were being collected.

This book is the result of this multimethod research project and personal journey in the understanding of what it means for families to live at a historical time when everything we do leaves a data trace and this data trace is than collected, stored, and used to define us as data subjects. During my research I met a variety of parents that—unlike Carlos—were not worried about the surveillance and processing of their personal data. They confronted me with the same questions: "So what? I have got nothing to hide?"; "Why does it matter if companies can track what I buy online to personalize products or send me targeted advertising?"; "Isn't that good?" As this book will show, it matters. It matters because today the boundary between consumer data and citizen data has become increasingly blurred. It matters because our personal data is used to make data-driven decisions that impact our lives and the lives of our children in all sorts of ways.

SURVEILLANCE CAPITALISM AND THE DATAFICATION OF EVERYTHING

Over the last decade or so we witnessed a "revolution"[3] of a sort, one that has radically transformed the ways in which we relate to data. This understanding—if stripped from techno-deterministic perspectives—sheds light on some of the crucial transformations of the last decade. In the last ten years, we have become the first witnesses of a radical technological transformation. We have not only seen the rise of supercomputers able to integrate larger and larger data sets, but we have also witnessed key developments in artificial intelligence. Yet, when we think about this transformation, we cannot only focus on technological change, we also need to consider that governing institutions, educational bodies, health care providers, businesses of all kinds, and multiple other agents have started to turn every aspect of everyday life into data (Mayer-Schönberger and Cuckier 2013) and are using this data to make key decisions about our lives (Kitchin 2014; Lohr 2015; O'Neil 2016).

If we want to really understand this transformation, we need to look at a change in our economies and consider the rise *surveillance capitalism*. According to Bellamy-Foster and McChesney (2014), surveillance capitalism established itself over the last decades as a political economic system defined by the relations of power between governments, military powers, secret agencies, the financial sector, advertisers, internet monopolies, and multiple other agents who surveilled, controlled, and capitalized on individual data (Bellamy-Foster and McChesney 2014). Zuboff (2015, 2019) brought their argument further and explored the ever-growing networked infrastructure of *surveillance capitalism* by considering the role played by companies like Google, Amazon, and Facebook, which constantly sought new ways to turn personal data into profit. Zuboff (2019) argued that Google played a fundamental role in the emergence of surveillance capitalism, which was very similar to the role that the Ford Motor Company and General Motors played in industrial capitalism. In fact, Google has not only radically transformed the economy but also society as we know it.

Under surveillance capitalism, companies like Google and Facebook make profit by transforming everyday human experience into behavioral data, which is then used for targeted advertising. This data-driven logic has permeated businesses, institutions, and law enforcement alike, who have bought into the promise of big data and AI systems. The promise (and value) of big data and AI lies precisely in predictive analytics, in the understanding that the aggregation of data can highlight behavioral patterns, which then can enable companies to somehow predict the future and to mitigate risk (Lohr 2015).

Although it has become clear that algorithmic predictions are often biased, discriminatory, and wrong (O'Neil 2016; Noble 2018) in the age of surveillance capitalism, predictive analytics is used everywhere—by educators in schools who believe in creating personalized education and by banks, insurers, and recruiters who need to decide loans, premiums, or whether one is a good fit for a job or not. Predictive analytics is also used by the police (Dencik et al. 2017), by immigration enforcement, and by governmental institutions who decide a variety of issues from child protection to social welfare (Eubanks 2018), and of course by secret services.

THE DATA SUBJECT AS CONSUMER AND CITIZEN

One of the most deeply problematic aspects of surveillance capitalism, as Carlos noticed, lies not only in the fact that our data is constantly surveilled and monetized, but also that predictive analytics can only work if companies use our data traces to profile us and construct us as data subjects. As data subjects, we are no longer only consumers making choices, but we are consumer subjects that have been sorted by our own choices (Ruppert et al. 2017). This becomes clear if we look at the role of data brokers. According to the Federal Trade Commission (FTC), the data collected by data brokers relates to numerous different dimensions of an individual life:

> Data brokers collect data from commercial, government, and other publicly available sources. Data collected could include bankruptcy information, voting registration, consumer purchase data, web browsing activities, warranty registrations, and other details of consumers' everyday interactions. Data brokers do not obtain this data directly from consumers, and consumers are thus largely unaware that data brokers are collecting and using this information. While each data broker source may provide only a few data elements about a consumer's activities, data brokers can put all of these data elements together to form a more detailed composite of the consumer's life. (Federal Trade Commission 2014, IV)

The report highlighted the lack of transparency and accountability of data brokers and raised fundamental concerns about the multiple ways in which data was being used to profile consumers and how these constructions may impact their life. This is evident in the following:

> Marketers could even use the seemingly innocuous inferences about consumers in ways that raise concerns. For example, while a data broker could infer that a consumer belongs in a data segment for "Biker Enthusiasts," which would allow a motorcycle dealership to offer the consumer coupons, an insurance company using that same segment might infer that the consumer engages in risky behavior. Similarly, while data brokers have a data category for "Diabetes Interest" that manufacturer of sugar-free products could use to offer product discounts, an insurance company could use that same category to classify a consumer as higher risk. (Federal Trade Commission 2014, v)

What I find interesting about the FTC report is that individuals are identified as consumers. Yet when the data collected and profiled—as the report suggests—is about voting registration or other key details of an individual's life (which of course would include religion, ethnicity, class, gender, politics, and so forth) then we are not simply talking about consumers and consumers' rights, but we are talking about citizens and citizens' rights.

In the emerging infrastructure of surveillance capitalism, as described by Zuboff (2019), the data subject becomes simultaneously a consumer subject and a *citizen subject*. Today, individuals often give their consent to share their consumer choices with companies and advertisers, but this data is often used for political and civic purposes to profile individuals as citizen subjects. This latter point is becoming increasingly more evident. In 2018, for instance, the Cambridge Analytica/Facebook scandal went viral. The scandal was first highlighted in 2015 by an article that appeared in *The Guardian* (Davies 2015), which suggested that republican candidates in the US were using a little-known company, Cambridge Analytica, who was able to harness the sensitive data of tens of millions of Facebook users to profile them politically and influence them with targeted content.

It turned out that Cambridge Analytica had access to that data, because a researcher Aleksandr Kogan, created a very standard Facebook app that was able to harness not only the data of the people who used that app, but also the data of their friends and networks. Through 270,000 people who opted in, Kogan was able to get access to the data of some 50 million Facebook users and to pass that information on to Cambridge Analytica, who was able to build psychometric profiles of users (Meyer 2018).

The 2018 Cambridge Analytica/Facebook scandal was perhaps the very first example in the public domain of the multiple ways in which data brokering and digital profiling is not only impacting our consumers' rights (e.g., determining the price we get for things, or the access we have to specific products such as insurances), it is impacting our civic and democratic rights (e.g., impacting the right to express our political views without being judged and framed). What the Cambridge Analytica scandal revealed was that individuals were being profiled for their political views and sent specific information aimed at having an emotional impact and influencing their electoral choices.

The Cambridge Analytica scandal is one of the multiple examples where the line between the consumer and the citizen subject becomes blurred. At present we are surrounded by a variety of examples. One of the most well-known examples in the US is represented by Amazon, who has started to sell facial recognition software to law enforcement and border police (Statt 2018). In the UK in 2018, the Labor Party bought the data of new mothers or mothers to be to politically profile them (Pegg 2018).

It is precisely because the line between citizen/consumer data is becoming so blurred that at present we are seeing the emergence of new public debates, lawsuits, and regulations that are aimed at addressing this societal transformation. In the EU, the implementation of the General Data Protection Regulation in May 2018 is holding businesses, organizations, and institutions accountable for protecting individual privacy. In the US, we witnessed the first lawsuits that address the problem of inaccurate profiling and data harm. Despite these changes, research on the complex relationship between data and citizenship is just emerging and we are only starting to address the fact that we are all being turned into *datafied citizens*.

When parents ask me why data privacy matters, my answer is that it matters. It matters because in the age of surveillance capitalism the "data subject" is much more than an individual consumer or a user sorted by one's own choices, it is a citizen whose freedoms and rights are highly dependent on the ways in which she is being profiled.

THE DATAFICATION OF CITIZENS

This book will argue that one of the main transformations brought about by surveillance capitalism is that *data traces are made to speak for and about us* in public in ways that were not possible before. Hence we are witnessing the rise of a new type of public self, the *datafied citizen*. In contrast to the digital citizen, who uses online technologies (and especially social media) to self-construct in public (Isin and Ruppert 2015), the datafied citizen is defined by the narratives produced through the processing of data traces; it is the product of practices of data inference and digital profiling (Barassi 2016, 2017; Hintz et al. 2017, 2018).

The datafied citizen is often governed by what Cheney-Lippold (2017) described as *ius algorithmi*, the law of algorithms. According to him, the law of algorithms is similar to other types of laws that control citizenship (e.g., *ius soli*, for which citizenship is granted on the basis of birth territory, or *ius sanguinis*, for which citizenship is granted on the basis of hereditary blood ties). Although *ius algorithmi* may not provide individuals with passports, it determines what rights they have access to on the basis of their datafied behavior.

When we think about the emergence of the datafied citizen, we of course need to take into account that the governing of citizens on the basis of their personal data has a long history, which can be dated back to the census of ancient civilizations (Hintz et al. 2018) or the establishment of the modern nation state (Lyon 2001, 2007). Yet we cannot be blind to the fact that, with the rise of surveillance capitalism, we are witnessing a completely new situation. We live in a world where a plurality of agents can cross-reference large amounts of our personal data and profile us in often obscure ways. They use the data that we produce—and the data that others produce about us—to track us throughout our lives so that they can find out our behavioral patterns. With this data they make assumptions about our psychological tendencies and construct narratives about who we are. As datafied citizens, we do not have any control over the narratives produced through digital profiling, even when these narratives are discriminatory and wrong (Gangadharan 2012, 2015; Eubanks 2018; Noble 2018).

This book is based on the understanding that if we really want to explore the emergence of the datafied citizen, we need to look at children. Children are key to the understanding of how citizenship is being transformed by our data-driven cultures. They are the very first generation of citizens who are being datafied from before they are born. From the moment a child is conceived, important personal data is uploaded on social media or pregnancy apps. As children grow up, most of their health and educational data is stored and processed by data brokers and AI technologies. All these different forms of data monitoring and processing are just the tip of the iceberg. The picture becomes much more complex if we consider the role played by home hubs, facial recognition technologies, and voice prints. What is becoming obvious is that children's personal information is now being collected, archived, sold, and aggregated into unique ID profiles that can follow them across a lifetime.

Children have traditionally been excluded from debates about citizenship and public life (Nolas et al. 2016). In western liberal thought, they have always been stripped from their autonomy and agency by subordinating them to adults and describing them as noncitizens or future citizens (Coady 2009). However, in the last thirty years or so—since the founding of the United Nations Convention on the Rights of the Child—there has been a move from an understanding of children as subordinated to adults and needing protection, to that of appreciating that children were self-determined citizens with specific rights (Earls 2011). Yet also in the discourse about children's rights, children's needs and interests are often not taken into consideration, especially in relationship to data privacy rights (Livingstone and Third 2018).

My understanding is that the new data environments we live in are exacerbating the lack of agency and autonomy of children in defining themselves as citizen subjects. Although digital environments have provided children and youth with the possibility to perform their public selves online (Third and Collin 2016), under surveillance capitalism, children's data traces speak for and about their life trajectory in ways that were not possible before.

Children are stripped from their autonomy not only because parental consent is used for gaining lawful rights to process children's data but also because children's data traces are produced, shared, collected, and processed by others beyond parental consent and control. In different sectors of their everyday life from education (Williamson 2017a, 2017b) to entertainment and health, children's data traces are used to find out their behavioral patterns, make assumptions about their psychological and behavioral tendencies, and construct narratives about who they are.

These narratives are often based on the social context in which they grow up in. As Taylor (2017) argued, data technologies often sort, profile, and inform action based on group rather than individual characteristics and behavior (ethnicity, class, family, and so forth). Therefore, children are being profiled on the basis of the families and the social groups they belong to. Children have historically been profiled on the basis of the families and the social groups they belong to, but in the past—once they grew up—they could choose to become their own persons and separate themselves from the values and beliefs of their family.

The difference between me and my daughters is that there's very little information about my childhood out there. There's no record of whether my mom smoked when she was pregnant, whether any members of my family have a criminal record, or of all the embarrassing things I did when I was a teenager. I could grow up to be my own person, to choose my own values and live by them. My children won't have that liberty—for the rest of their lives, they may be judged by the data points collected through their childhood. The way in which they are profiled today might follow them throughout their life-span.

Turning children into data subjects could stall social mobility (Eubanks 2018) and impact their right to self-definition and moral autonomy (Nissenbaum 2010). Therefore, when we think about the datafication of childhood, critical questions are emerging about the ways in which children are being datafied into citizen subjects from before they are born. Looking at the datafication of childhood can enable us to start asking and tackling critical questions about our democratic futures and the type of society that we are building.

THE CHILD | DATA | CITIZEN BOOK

In this book I will be exploring the datafication of children by looking at the practices and the choices of others around them: the parents, the school headmaster, the doctor, the relative, the friend. I will also explore the business models and strategies that make the datafication of children possible. I will not focus on children as such but on their data flows and traces. This implies that in thinking about the datafication of childhood, in this book the child is very rarely an agent. I am aware of how problematic this position is, especially at a time when scholars working on childhood are joining efforts to highlight and emphasize the *agency of children* in our societies (Oswell 2013).

All the literature on children's agency is pivotal and necessary, and this is especially true if we turn our attention to digital technologies. There is outstanding research that focuses on children's digital practices. The following projects and research hubs are an excellent example of this: EU Kids Online, Toddlers and Tablets, Net Children Go Mobile, the Connected Learning Research Network, Children's Digital Media Center. All these different and interconnected research projects show us that, when we think about technologies, children cannot be lumped together with adults as part of the wider population, but their digital experiences, needs, and concerns matter in their own right (Livingstone 2009). At present there is also excellent research on the civic agency of children (Nolas et al. 2016) and how they understand data privacy and negotiate with surveillance capitalism (Stoilova et al. 2019).

Although I recognize the importance of research on children, in this book I will not focus on their lived experience, beliefs, and practices. Here the child is not really a child or an agent, it is analyzed as a *discursive* subject, a construct that is socially produced and context specific (Wyness 2011) by different parents, different technological platforms, and different data regulations. What I do in this book is to shed light on the everyday practices, beliefs, policies, political economic systems, business models, and technologies that are making the datafication of children possible.

This book will show that the datafication of children is not a linear, cohesive, or even a rational process that is transforming them into quantifiable data subjects, but it is a rather complex and messy process that is defined by a plurality of technological possibilities, designs, and organizational intentions. These highly diverse forms of data collection combined with the unpredictability of human practices is leading to the creation of large amounts of children's personal data that is often fragmented, imprecise, and mistaken. In this context, we need to start tackling serious questions not only about the ways in which children's data is surveilled and used, but also about the ways in which children may be constructed as data subjects on the basis of very inaccurate data traces, which—as Carlos mentioned—do not reflect real life.

In its exploration of the datafication of children, the book will be divided in three different parts: *the Child, the Data, the Citizen*. The first part of the book (The Child: two chapters) will explore how the social world of the family is becoming datafied under surveillance capitalism, and this is leading to the everyday production of large amounts of children's data. This part will engage with critical debates about techno-dependency, parental consent, and data-tracking in family life. The second part of the book (The Data: four chapters) will discuss four different types of children's data flows: health, educational, home life, and social media. This section will combine ethnographic findings with an analysis of business models and data policies. In this part of the book I will argue for the importance of moving beyond understandings of *personal data* as a unique umbrella term and to actually reflect on the different typologies of data that are collected from family life, as well as on their different impact. In the third part of the book (The Citizen: two chapters) I will critically reflect on the complex relationship between children's data flows and the datafication of citizens from birth. These last two chapters will be largely theoretical and will combine an analysis of the latest research in the field of data privacy and AI ethics with theories of personhood and digital citizenship. I will discuss what it means to think of children as datafied citizens and will engage with key questions about individual rights, algorithmic inaccuracy, and data justice.

This book was in production when the pandemic struck, so it does not include a thorough reflection of the COVID-19 crisis. Yet, following the pandemic, the processes of datafication of family life that are described in this book have accelerated and intensified. Understanding these processes, questioning the role of Big Tech companies, and investigating our society's attitudes to personal data will be the top priority of our times, if we want to shed some light on our uncertain futures.

1

DIGITAL ROUTINES: HYPERCONNECTIVITY AND FAMILY LIFE

"I pick up the phone so many times in a day, and I look for things, constantly," Mike told me when I interviewed him. Mike, who lives in Los Angeles, is the father of two children aged 5 and 12 years. He paused, looked up, and then added: "Don't ask me what I am looking for, because I really don't know." Like Mike, I also feel that I am constantly picking up my phone to "check" for things. Many times, this need for constant updates and connectedness is completely unmotivated, and I am not sure what I am looking for when I look at my phone. Claire, who lives in London and is the mother of two children under five, described this feeling as an "uncontrolled compulsion." One day she told me that she found herself "scrolling Facebook" while her children were having breakfast, and suddenly she realized that she: "Could not tell why she had picked up the phone in the first place."

If we want to understand the datafication of childhood, we need to start by considering the pervasiveness and inescapability of technology use in family life, especially for those families living in high-tech cities like London and Los Angeles. Today, daily routines are shaped by digital use. From the moment parents wake up and reach for the phone, to the moment in which they find themselves, in the evenings, sitting on the sofa "double screening." Daily life has become more and more defined by digital routines.

This was not always the case. All the parents I worked with could remember the time in which, "You didn't always *have to be* connected"; a time in which "You chose to go online." In describing this change, many parents like Claire referred to ideas of "compulsion" or talked about "technological addiction"; they blamed themselves and their partners for what they perceived as excessive technology use, and bought the latest apps to curb their alleged "techno-dependency."

If we really want to understand hyperconnectivity in family life and make sense of how daily routines have changed, we need to move beyond the narrative of individual techno-dependency and stop blaming parents for their technology use. What we need to do is to critically reflect on our data economies and shed light on the multiple ways in which surveillance capitalism—through the datafication of services and the creation of technologies such as "infinite scrolling" that are designed to promote compulsive behaviors (Andersson 2018)—is promoting a new time regime; one that pushes users to produce more and more data and generate more income and value.

DIGITAL ROUTINES

> I reach for my phone the second I wake up. I might say "hi" to Scott, but I take the phone into the bathroom to go peeing. I just sit there and wait to wake up with my phone. I feel that this is how my eyes focus. (Nicole—Los Angeles, two children, aged 8 and 10 years).

> The minute I wake up I look at my phone. I check my emails, my Instagram, my Facebook, and then I would go on the news; it is pretty much always in that order. Depending on how long it has taken me, sometimes I do the cycle again, because new things might have happened. (Sonia—London, one 4-month-old baby)

Over and over again as I talked to parents, I asked them to describe their digital routines. I deliberately decided to ask parents to describe their digital routines not on days when they had to go to work, but on weekends or holidays, when they could disconnect and focus on family life. There was an extraordinary overlap and similarity of experiences, and it did not matter whether parents, like Sonia and Nicole, came from very different class and cultural backgrounds or they lived in Los Angeles or London.

From the moment they woke up, both Sonia and Nicole (as well as a variety of other parents that I worked with) reached out for their phones. That would start a day of constant connectedness. At breakfast, Nicole's girls would watch television while she "scrolled through her phone" and in Sonia's house: "The TV is always on, and sometimes the laptop as well; we sit on the sofa, and we have the phone next to us, and we always pick it up and put it down, we are always looking at it. We would talk, watch the phone or the TV and we would play with the baby."

Sonia and Nicole both mentioned that once they left the house, if they were going somewhere by car or public transport, they would use their phones and other technologies to check where they were going, to communicate with people, and of course to entertain the kids. Nicole's girls would have cartoons in the car, and Sonia's baby would watch sensory light videos on YouTube in his stroller or would have the "white noise" app turned on when he was sleeping.

During the day, as they participated to different activities, visited friends or took the children to the park, they would take pictures and post them on Facebook and Instagram. They would then fall into the "feedback loop" and constantly check for new notifications. They would also check the news regularly. Throughout the day, no matter what they were doing, they would pick up their phones at least every 15 minutes. In the evening, they would watch a movie and "double screen" on their phone.

Sonia's and Nicole's experiences are of course not isolated. Today, parents often use their mobile phones and other technologies to connect to others through emails, messages, notifications; they use them to plan their days ahead, "look for stuff to do," and "check if Amazon deliveries have arrived"; they also use them to find out about news and different types of information, or simply to create a "feeling of home" through music, films, and other forms of entertainment.

Two anthropologists in the UK describe the inextricable relationship between technology use, everyday routines, and "making the home feel right" (Pink and Leder Mackley 2013) and reveal how humans today very rarely disconnect from their technologies, which are often left on standby. In the lives of parents like Sonia and Nicole, and in my own, technologies are always there on standby. They lie on side tables, on kitchen counters, in cars and purses; they follow us and coexist with us and constantly seem to be waiting to be picked up, checked, and engaged with. They have come to define our daily routines.

When Sonia and Nicole described their digital routines, they told me that they could remember a time when their relationship to technologies was very different. Nicole told me that: "Only ten years ago you didn't need to be online all the time. You didn't have to wake up in the morning and go online." Sonia, instead, told me that in the past she could "decide" to go online, while today is totally different:

> I bought my iPhone in 2009, at the time I wasn't as connected as I am now, it is integrated into life now. Before you had to make a decision if you wanted to go on the internet, now you are constantly connected.... Now I check my phone constantly, if I wasn't sitting here with you I would be on my phone. I try to make a conscious effort not to do it, but I can't tell you how much I look at my phone.

When I asked parents if they could remember their past digital routines, they often would mention—like Nicole and Sonia—a radical change in their technology use. Luke, the father of two children under five, who lived in West London and was an entrepreneur, told me that in the late 1990s and beginning of 2000s he would go online once a week, while today he checks his phone every 10 minutes. At the time he was still living at home with his two brothers, who would constantly argue over access to the computer. Sipping his beer in a noisy pub of central London, he reminisced about his early relationship with online technologies and the arguments with his brothers. "Do you remember the sound of the dial-up connection; the screechy noise?" he asked. I did. Luke looked bewildered and amused. He then added: "I would go online once a week? Can you imagine that?" No, I couldn't.

Of all the stories I heard from parents about technological change, Amy was perhaps the most lucid and poetic. I had met Amy in a park in Los Angeles in early 2016. Amy worked as a nanny and had a 5-year-old daughter. She agreed to sit down for an interview in early 2017, and we had a long chat over lunch. Amy told me that she felt that she was "forced" to transform her relationship with digital technologies, because technologies had become more and more pervasive, and she did not have a choice:

> A: Everything started to change around me. The world was changing. It was not me keeping up with the world. I grew up in this area where at first there were fields, and a house across the street and a pond and ducks. That's it. And then within 10 years there were all these houses, and my sister and I would joke about it, and say the houses would just sprout. I feel that the same thing happened to technology, every app spurred another app, it was an exponential growth. There was no way to stop it.

Amy complained about society's obsession with technologies; she believed that we had all become dependent on technology and that there was no escape. Also Nicole and Sonia talked about *addiction* and *dependency*. Nicole described her hyperconnectedness as a "need" that she could not control, and Sonia told me that she tried to resist the compulsion of using her phone, because she felt that she had "to live a bit more," but that she failed. The narrative of techno-dependency was widely shared among parents and was a key definer of their biographical accounts of technological change.

TECHNO-ADDICTION IN FAMILY LIFE?

One day, I had the pleasure to sit down with Jen, the mother of two children aged 3 and 6 years, who was a corporate lawyer and lived in Los Angeles. When I asked her to describe whether she experienced a change in her technology use over the last decade, she answered:

> J: When I got the BlackBerry I became so dependent on the flashing light. Whenever the light changed from green to red I would check it, and that was traumatic because I hated it. I worked for a law firm who told us that we needed to be available 24/7 and we needed to sleep with our phones. I don't remember much of the change with the iPhone. But, if I have to be honest it was different from today. At the time I was not so much dependent on going online, not even after I got the iPhone, now whenever I have a spare moment I go online. I am more dependent on this now than I was then.

Jen was married to Carlos who I introduced in the introduction to this book. She told me that she and Carlos were concerned that their techno-dependency would interfere on their children's life, and tried to "stay clear" from the phone, especially in the hours in which the children were at home and awake. She then added: "We try to be conscious about our phone use. To say that we are successful at it, it's a lie. But we try to be as conscious as possible."

I could immediately sympathize with Jen and Carlos, because I often struggled with my phone use, and I would constantly fail to contain it, even when I tried. During the Child | Data | Citizen project, I documented in a diary all those personal instances in which I—like Jen—felt that I could not stop myself from checking my mobile. I clearly remember one particular episode. In June 2017, I was in the car with my family in Los Angeles. The traffic was unbearable, as always. I was 8 months pregnant and tired. My eldest was looking outside her window playing "I spy with my little eye ..." I was sitting next to Paul, my husband, playing with my daughter and chatting with him. I told him how proud I was that—unlike other children who are constantly digitally entertained—our daughter would play on her own and was interested in the world. We continued our game. Yet I struggled to keep my hands off my phone. I must have looked at it at least ten times in a twenty-minute drive. What did I check? Oh, multiple things. I checked my emails to see if there was news from a new preschool that I had just contacted. I went on Twitter to see if anyone reacted to my tweet. I read the UK news and worried about Brexit. I wrote a couple of texts to make plans for the weekend, and I checked what bassinet I should buy for my baby. I did all that, while I played "I spy" with P.

As I was writing my fieldnotes I reminded myself of the book *Alone Together* by Turkle (2011) in which she describes her own compulsion and lack of control when it came to her digital use. She also talked about the fact that digital technologies make you feel pressed for time because people use them to multitask and to be always on. In the book *The Parent App* (Schofield Clark 2013), beautifully analyzes how families experience the time of technologies. She shows that parents rely on technologies to organize their complex work–life routines yet struggle with the imperatives of constant connectedness.

Exactly as they describe in their books, I also feel that I am constantly hyperconnected and multitasking, and I feel the pressure. My hyperconnectedness, as well as Paul's, triggers constant tensions in day-to-day life. I blame him for being always on his phone, and he blames me for the same. My eldest daughter often objects to our phone use, and our toddler recognizes our phones. When she sees them lying about, she picks them up and rushes to give them to us, almost alarmed as if they were the most important things, shouting "Mamma" or "Daddy." I always feel awful when she does that. My family's techno-related tensions are not uncommon; technological interference or *technoference*, to use the term by McDaniel and Radesky (2018), has become the new norm in family life, and this is giving rise to multiple tensions within family life.

TECHNOFERENCE? DAILY TENSIONS AND INDUSTRY SOLUTIONS

Sociologists and researchers working on children and digital media, have shown that "screen time" has become a contested terrain of negotiation between parents and children (Livingstone and Franklin 2018). Parents find themselves regulating the amount of media exposure that children are allowed to have. They worry, sometimes excessively, about screen time without critically reflecting on other issues such as how children learn through technology use (Livingstone and Blum-Ross, 2017). In the families I worked with, the issue of screen time was a big problem for many parents. The parents I interviewed were concerned about their children's techno-dependency or "obsession" with screens. One day, Alice, a mother of two children under ten who lived in Los Angeles, told me that she felt that her children were "too dependent" on their devices, and then she added: "Whenever I tell them to put them down, and go and play, they would count the minutes until they are allowed to connect again."

Screen time was certainly an issue for the families I worked with. Yet, my research revealed that key tensions have emerged not only between parents and children, but also between co-parents, who complain about their partners' technological addiction. Katie, who lived in Los Angeles and had a 6-month-old baby and a 13-year-old son, clearly experienced this tension. One day she told me that her fiancé was "on the phone all the time" and made her feel extremely frustrated, especially because he was always telling off her older son for being on the phone, while he was doing the same.

Most of the parents that I interviewed believed that technology use interferes with everyday relationships in family life. In order to make sense of this interference, parents often rely on the narrative of techno-addiction. Mike, who I mentioned earlier, was married to Zoe and felt overwhelmed by his wife's hyperconnectivity:

> M: I wake up in the morning with Zoe's phone not mine, she takes it in the bedroom and puts it next to her nightstand. I do not. I leave it in the other room. I need to disconnect. ... Zoe actually checks her emails and texts as she goes to sleep, and if she receives one in the middle of the night I have heard her waking up and checking her phone, and the first thing she does in the morning is to check her phone. I think she has a problem. You know if she can't find it or if she is separated from it for a period of time she has withdrawal symptoms, there is a physical reaction. ...

During the research I met a wide variety of parents, like Mike, who complained about their partners' technology use. Lara, a mother of four children aged 9 to 21 years and a grandmother of two, who lived in Los Angeles and worked as a nurse, struggled with her husband's mobile phone use. One day she told me that she felt that technological addiction was impacting their relationship:

> L: You know you need to be really disciplined if you are not careful enough it [mobile use] could destroy families, it could destroy your

> spiritual relationship, because it keeps you away from that. I noticed it in my relationship with God. I used to get up and read the Bible but then I entered a stage in which I woke up and started looking at my phone, so I had to discipline myself. I notice this with my husband. If it's not the TV, it's the phone, and we can't go to the store without him not putting his phone away. You know I tell him: look at me, we used to be able to spend time together.

The parents I worked with, often blamed themselves and their partners because they felt that their heavy phone use impacted their family life and their relationship with their children. Their worry is not totally unfounded. According McDaniel and Radesky (2018), parents' heavy digital technology use can have an impact on children's behaviors, especially when parents use these technologies as a way to mitigate stress. In those instances, children may act up to look for attention.

I found the article interesting because it introduces the term *technoference*, to describe how technologies can interfere with meaningful interactions between parents and children. Although there is not enough research out there to draw conclusions on how parents' technology use really impacts children, during my research—and in my everyday life—I have been exposed to so many different examples of technoference. One day in particular, when A was six months old and I was still on maternity leave, I spent a long time on my phone trying to sort out my photo gallery. When I looked up, I saw her trying to make eye contact with me and smiling. I felt terrible and wondered how long she had been trying to connect with me, and how long I had been distracted.

Many of the parents I talked to, like me, felt guilty about their phone use and they actively sought solutions to what they perceived as an "addiction." They talked about phone use as something that was "their fault"; as something that they needed to "cure" and "control." Each family had its own tactics to control phone habits, and these tactics varied enormously: some would delete social media apps, some would set specific rules, others would hide their phones. Among all the different practices adopted by parents to control their phone use, it was fascinating and paradoxical that at times parents bought apps that were specifically designed to "curb phone use" either to be more productive or to fight addiction.

Scott, who was Nicole's husband strongly believed that his phone use impacted the well-being of his daughters, so he downloaded the Forest App on his phone to control his phone habits. The app enabled Scott to regulate his phone usage by blocking his phone for 25 minutes. Once the phone was blocked, an animated tree started to grow. If he completed the 25 minutes, then he would earn credits that were than used by the company to plant real trees around the world. But if he picked up his phone while the tree was growing, then it would wither. During our interview, Scott showed me how the app worked, and when he picked up the phone, he told me "look it died; it's really impactful, it really makes you feel guilty."

We are living in a historical time when, paradoxically, parents are buying new mobile apps that are designed to control so-called "phone-addiction." At times these apps emphasize the narrative of technological addiction in their promotional blurbs. Understandings of techno-addiction are widespread in mainstream media and popular culture. At present there seems to be a shared assumption that because technology use (and especially gaming and social media) increases dopamine levels and pleasure circuits, it can trigger addiction.

Earlier works on social media addiction seemed to suggest that there was a connection between online gratifications and dependency in users, especially if we consider gratifications as information seeking, personal status, or relationship maintenance (Song et al. 2004); others have looked into online gaming and addiction (Kuss and Griffiths 2012). Yet technological addiction seems to be incredibly complex to understand and prove. In 2016, Andreassen et al. (2016) surveyed 23,533 adults and concluded that technologically addictive behaviors varied according to different users, and that the concept of "internet use disorder" was not warranted.

Ferguson (2018), a professor of psychology who researches technology use and families, recently debunked the myth that technology use can be compared to substance abuse. He argued that technology use causes dopamine release similar to other normal and fun activities, which is about 50 to 100 percent above normal levels. A substance like cocaine, in contrast, increases dopamine by 350 percent. Ferguson (2018) believes that people who claim that brain responses to technologies and drugs are similar are trying to liken the drip of a faucet to a waterfall.

The understanding that technological addiction is not a given fact of our everyday life, and that it is very difficult to prove, makes us realize that perhaps we should avoid looking for a psychological cause to explain hyperconnectivity and technoference in family life. Rather, we need to look at the broader picture and consider how hyperconnectivity is enhanced by the very business models and technological designs of surveillance capitalism.

ADDICTIVE TECHNOLOGIES AND TIME REGIMES

In November 2017, during a talk at Stanford University, a former vice president at Facebook equated social media—with their dopamine-driven feedback loops—to cocaine, and he said that he felt a "tremendous guilt" about growing the social network (Wang 2017). In July 2018, the BBC's *Panorama*, a British investigative current affairs program investigated how tech giants make social media apps deliberately addictive. In the program, they focused in particular on the example of "infinite scroll" and interviewed its engineer, Mr. Raskin, who had worked for Mozilla and Jawbone. Infinite scroll allows users to endlessly swipe down through content without clicking, and by doing so it favors addictive practices (Andersson 2018). What was emerging from these new debates in the press was that computer engineers and the gurus of the Silicon Valley were openly admitting that the technologies they designed were in fact designed to make sure users were hooked and spent the most time possible using them.

While the mainstream media immediately jumped to quick conclusions about the relationship between technological design and addiction, Professor Andrew Przybylski, an experimental psychologist and Director of Research at the Oxford Internet Institute, together with other psychologists cautioned against simplified understandings of psychology and addiction. In an article published in *Business Insider* (Hamilton 2018), they argued, like Ferguson (2018) had done, that technologies cannot really be compared to drug or alcohol abuse, and that associating them could be risky. They also argued that the computer engineers, like Mr. Raskin, who claimed to know much about psychological processes, relied on psychological research that had been unproven or criticized.

Paradoxically I became aware of the debate surrounding infinite scrolling while I was scrolling through my Twitter feed. As I was reading about the debate, I immediately thought that if we moved away from simplistic understandings of technological addiction, we would realize that what Mr. Raskin was talking about was actually not that new. Mr. Raskin claimed that computer engineers were designing technologies to influence how users spent their time in order to create value. The more time users spent on their devices, the more data they would generate, which in turn would create more value for the digital economy. Hence it was essential, from the perspective of developers, to design technologies that would guarantee some form of dependence or the illusion of it.

The practice of controlling people's temporal behaviors through technology use in order to create more value may seem absurd, but it has defined the history of capitalism. Since the very early days of capitalism, technologies have always been used to control people's temporal practices and create more economic value. This becomes clear if we take a step back in history. In the Middle Ages in England, for instance, there were only few areas of exact time keeping, such as monasteries or towns, and time was still organized around specific agricultural or social activities. However, from the fourteenth century onward with the protestant ethics and the rise of earlier forms of capitalism, the "time of the merchant" gradually took hold over other dimensions of time, with clocks entering households and taking over church bells in organizing the everyday lives of people (Thompson 1967, 82–86). This led to a gradual diffusion of a new type of "time-consciousness" (Thrift 1990), in which "clock time" was established—especially thanks to new technologies of time keeping—as the *hegemonic* form of time measurement that radically transformed the digital routines of the family. This process was enabled through institutions such as the factory or the school, which used incentives, fines, and other strategies to transform people's temporal behaviors (Thompson 1967).

The synchronization of human activities and clock time were all necessary for the economy of the modern, industrial nation state. This is clear if one looks into the key contributions in social theory. For Marx (1990), clock time was introduced to organize and manage labor time under capitalism. After the industrial revolution, time had become a valuable commodity, because the "management of time" maximized production and capitalist accumulation (Marx 1990). For Weber (1978), clock time enabled an increased synchronization of human activities, which was key to the rationalization process of the modern nation state. Without getting too theoretical, it suffices to say that time management and time regulation were a fundamental aspect of Fordism in the earlier days of the industrial revolution, and this type of time management was made possible thanks to the clock.

The 1970s and 1980s gave rise to a new type of economy, one that no longer relied on the manufacturing industry, but instead created wealth and value from the service industry. The new economy was fueled by a deregulated financial sector and the possibilities offered by the globalization of markets. It was also fueled by the new networks of communication, defined first by the development in satellite and broadcast technologies and later internet technologies. There are countless sociological and theoretical accounts of this sociotechnical transformation. A few examples of these theoretical accounts can be found in the works of Harvey (1991), Giddens (1991), and Castells (1996, 1997). The new global economy, thanks to the rapid expansion of internet technologies, introduced a new type of temporal regime that guaranteed that working routines were no longer dictated by clock time like in the factory, but by a self-regulating flexibility, which eroded the boundary between labor time and leisure time (Gill and Pratt 2008; Gregg 2011).

This new time regime was again established through technology use. In fact, internet technologies have created the shared impression that we live in a continuous present, a *hyper now*, where past and future are subservient to the logic of the present and where everything happens immediately (Hassan, 2003, 2007, 2009). Immediacy is profoundly powerful for us because it carries two different meanings. On the one hand it serves to indicate the compression of space and a sense of "proximity" (from the Latin, *immediate*); on the other hand, it serves to specify the compression of time and the notion of "instantaneity." In noticing these two different meanings, Tomlinson (2007) argued that we need to understand immediacy as meaning both instantaneous contact and immediate fulfillment (Tomlinson 2007, 91).

With the rise of surveillance capitalism, immediacy came to have a new meaning. Online technologies did not only enable instantaneous communication and immediate fulfillment, but they also enabled the continuous production, storage, and processing of the new oil: personal data. In the age of surveillance capitalism, data technologies have introduced a new time regime that valorizes continuous productivity (Fuchs 2013) and buries any unproductive communicator (Elmer 2004). This is what Mr. Raskin was talking about. The developers of data technologies need to make sure that users keep on producing more and more data, because data generates value.

SURVEILLANCE CAPITALISM AND A NEW TIME REGIME

When I think about my own experience and the experiences of parents that I worked with, in the light of these theories I cannot fail but to conclude that perhaps we are not to blame, and that maybe hyperconnectivity in family life has less to say about our individual compulsions and more to say about what it means to raise families in the age of surveillance capitalism. Understandings of hyperconnectivity and techno-compulsion have always existed since the very early developments of the internet. Researchers in informatics and computer sciences have long focused on the idea of ubiquity (Berleur et al. 2007), as a way to describe not only the mobility and interconnectedness of technological platforms but also their pervasiveness and embeddedness in everyday life. Researchers have also long studied the opportunities, challenges, and tensions that hyperconnectivity poses for humans and organizations (Quan-Haase and Wellman 2006).

Yet in the age of surveillance capitalism something has changed, and hyperconnectivity has come to define our everyday life in ways that were not possible before. This is not only because the technologies that families use in day-to-day life, such as infinite scrolling have been designed to ensure that people are constantly connected and that users produce more and more data and generate more income and value. It is also because, in our data-driven economies, more and more services and business have become digitized and datafied; hence, parents do not have a choice and are forced to sort out problems, pay bills, and communicate with their doctors or with their children's school by using data technologies.

Angela, a single mother of an 8-year-old who lived in North London, tried to explain this transformation. "It's just that everything is online, I always have an email to write or something to check, and I try to control it and make time for my son, but it's not easy." Her comment was spot on. One of the main changes brought about by surveillance capitalism is that the institutions and businesses that individuals encounter in their everyday life (e.g., health providers, education institutions, banks, traveling, leisure activities, and local governments, among others) are increasingly relying on digitized and data-driven services, and as Angela said, "Everything is online."

Although ubiquity and hyperconnectivity have existed since the earlier developments of online technologies, today they have been exponentially transformed, because families have no choice *but* to constantly be connected. It has become *a way of life*. It is through constant connectedness that parents can go about their daily life and sort out daily activities and problems. Floridi (2015) believes that we need to abandon the idea that there is a separation between online and offline. In our data-driven economies, everything has become *onlife*. Online technologies are no longer just tools, they have become "environmental forces" that shape and transform our self-conception, our mutual interactions, and our interactions and understandings of reality (Floridi 2015). Everything is onlife, not simply online. It is for this reason that families are being datafied.

CONCLUSION

Family life is today dictated by a new time regime of hyperconnectivity; from the moment in which parents and children wake up to the moment they go to bed, digital technologies have become a fundamental dimension of their daily experience. The stories of the parents that I worked with show that this situation is relatively new for them and that they remember a time in which they did not "need to go online as soon as they woke up." These stories also show that there are deep-seated tensions that are emerging in family life as a consequence of heavy technology use, and parents often feel that technologies are "stealing time" away from meaningful family interaction. They also feel that they cannot control their technology use; they are "hooked" and "dependent" and this techno-dependency is impacting their children.

In this chapter I argued that focusing on "individual compulsion" overshadows the political and economic forces behind hyperconnectivity and techno-dependency in family life. The narrative of individual addiction is really not enough to explain this societal transformation. Technologies that we use today are designed to be addictive. If we want to understand hyperconnectivity in family life we need to take into account these technological designs, and we need to question their business models and the political economic contexts that define them.

Historically, technological developments have often led to the establishment of specific time regimes, which maximize economic value. Although the clock was used during industrial capitalism to maximize manufacturing production, internet technologies were instead used to create a new flexible time regime of work life that was key to the development of the global economy. Similarly, data technologies are designed and structured in a specific way that enables developers to guarantee that users produce more and more data and generate more value. Furthermore, like it happened at the time of the industrial revolution, when society (for example, the factory, the school, the markets) synchronized to meet the demands of capitalist production, all the society around us is transforming and being "datafied" precisely to meet the demands of surveillance capitalism. It is for this reason that families have moved from online routines to an onlife experience.

Reflecting on the past to understand the present can enable us to realize that maybe parents are not to blame for their hyperconnectivity. The fact that I am now writing while checking my Twitter notifications at 10 p.m., and the children are in bed, perhaps can be understood by referring to the broader picture. The broader picture, as the next chapter will show, can enable us to appreciate that it is through our everyday, hyperconnected, digital routines that children are being datafied.

2

DATAFIED FAMILIES: CHILDREN'S PRIVACY AND THE PROBLEM OF CONSENT

In 2018, I read a Facebook post that was shared by one of the parents' groups that I followed for the research. The post mentioned that among the Himba Tribe in Namibia, when a mother decided to get pregnant, she would go and sit under a tree to hear the song of her baby. Allegedly that was the day in which, for the Himba, the child was born. The post concluded that among the Himba, "The birth date of a child is counted not from when they are born, nor from when they are conceived but from the day that the child is a *thought in its mother's mind* ..." (Facebook post, shared 2018).

I found the post very touching, and I was not alone. The story of the Himba received more than 120,000 likes and numerous comments of parents mentioning how beautiful and heartwarming it was. Such a reaction was unsurprising. For those who reacted to the post on Facebook, the story mattered. It mattered because it spoke directly about the importance of preconception and that for many, deciding to have a baby is a key moment in one's life.

When I read the post about the Himba, I was immediately intrigued and decided to find out more. The post dated back to 2015 and originally appeared on Tumblr. Since then, it was shared extensively not only on Facebook but on a variety of parenting blogs and websites. I also realized that the post was a fake. I searched anthropological journals and other journals to find more information, but I was unsuccessful. Then I found out that the idea was taken from a book written by Sobonfu Somé (1999), a Burkina Faso author and teacher, who talked about the "preconception" rituals among the Dagara tribe in Burkina Faso. In the book she explains that among the Dagara, women engage in preconception rituals because they believe that the spirit of the child exists before conception. She does not seem to mention the date of birth. Hence the post—with its romanticized images of tribal women and their babies—was yet another example of stereotypical and inaccurate understandings of African tribes.

I wanted to start this chapter with this anecdote because I found the post's emphasis on the importance of preconception paradoxical. Under surveillance capitalism, the data tracking of children and family life can begin precisely the day a child becomes *a thought in a parent's mind*. Prospective parents often search Google to find information about how to get pregnant or they download fertility apps to track ovulation. From the moment in which they *think* or decide to conceive, they are immediately profiled for their interest, and their future children have the potential to become data subjects.

This chapter will focus on the datafication of children. It will show that children are being datafied from before conception for two interconnected reasons. On the one hand they are being datafied because parents rely on a variety of data-tracking technologies that produce, track, and share children's personal data. On the other hand, they are being datafied because under surveillance capitalism, the businesses, services, and institutions that families encounter in their everyday life (e.g., bank, doctor, school, supermarket, among others) have all become data driven.

This chapter will explore the thoughts, feelings, and understandings of parents as they try to make sense of the everyday datafication of their children and family life. It will show that for many today, it has become impossible to escape this process of datafication or to protect the privacy of their children. By showing that families cannot really choose or opt out of the datafication of their children, the chapter will argue that we need to start reflecting on the fact that surveillance capitalism depends on the *systematic coercion of digital participation*, which forces people to give up their personal data and to comply with data technologies (Barassi 2017a, 2019). This, I will show, raises critical questions about the very notion of consent, upon which much data protection regulation depends.

DATA TRACKING IN FAMILY LIFE

> Track your period, ovulation, symptoms, moods, and so much more in one beautiful app! With customized ovulation predictions based on your unique cycle, Ovia Fertility makes it fun and easy to manage your health and achieve your goals … ⋆Comprehensive Health Tracking⋆: + Period and cycle + Moods and physical symptoms + Intercourse + Nutrition and weight + Activity and exercise + Sleep + Ovulation and pregnancy tests + Cervical fluid + Medications + Blood pressure + Basal body temperature. (Ovia App, promotional blurb; Ovuline Inc. 2018)

The datafication of family life can begin from the moment in which a parent thinks about having a baby. Prospective parents, as mentioned previously, often use Google and multiple websites to search for information on how to conceive. They also use fertility apps, like Ovia, which enable the production and gathering of vast amounts of health data of the woman, such as information on intercourse, nutrition, body weight, blood pressure, moods, and symptoms.

After conception, many families download pregnancy apps. The market for these apps has grown enormously. In 2015, among 165,000 mobile health (mHealth) apps, 7 percent focused on women's health and pregnancy (Misra 2015); between 2013 and 2018, the WebMD pregnancy app was downloaded 1.6 million times (Scripps Research 2018). The market of pregnancy apps has grown exponentially because parents use them to track the health of the unborn baby and the health of the mother.

Tracking the unborn and women is certainly not new (Lupton 2013). Yet with the use of pregnancy apps, this surveillance and tracking, which has always occurred through medical practices and images, has reached a new dimension (Leaver 2015; Lupton and Thomas 2015; Barassi 2017a). In fact, the extensive use of pregnancy apps by parents is enabling a situation whereby corporations have access to important data of the unborn such as conception date, weight, number of kicks in the womb, possible names, cultural background, heart rate, diet before conception, parents' thoughts, family ties, family medical history, complications during pregnancy, and due date to name a few. These apps are thus the very first technologies through which families and children become datafied.

Once a baby is born, parents sometimes use baby trackers, or they use wearables to manage the baby routine and track sleep, feeds, and bowel movements. Again, there is nothing new in the tracking of the baby. Families of newborns have historically tracked this information in journals. When my first daughter was born, my mother showed me the journal that she kept of me as a newborn. Written in black ink on yellow pages and in my mother's familiar handwriting, there was a list of feeding times, naps, and diaper changes. She kept the journal in a drawer of her study and no one outside our family had access to it. Consequently, even if the tracking of the baby, like the tracking of the unborn, has always existed, baby apps—with their charts, reports, and interactive elements—have greatly transformed this historical practice and given it a new datacentric dimension.

One day for instance, I was interviewing Katie, who lived in Santa Monica, Los Angeles, with her 13-year-old son and a 6-month-old baby. Katie told me that data tracking was key to the "running of her family life," and she was particularly grateful for a baby tracking app:

K: So me, my husband, and the nanny are all connected to the same baby app, and we can all log in and see everything. And we can say, "Hey, here is pattern!" If one person comes in and takes over, you don't need to remember everything, and you can look at the data and you know: OK, he has napped at this time; he ate at this time, and so on.

V: Wow, that's handy.

K: Yes, and the app also gives you a full summary of the day, which I love, because I love analytics, so it tells me how much he ate, how much he pooped. I have it all right here. Then if I go to the doctor, I can say: here's his month's summary.

V: So, what do you love about that?

K: Data. I love data. So I don't have to remember. I have it all here. You know, an average of 8.5 poops per day (laughs). … But I love data when it comes to work. I love data when it comes to everything, because it gives you information and you can plan. I also use self-tracking apps for fitness for the same reasons.

In her enthusiasm for data tracking, Katie was not alone. Many of the parents I interviewed shared her view about the importance of data tracking, especially during pregnancy and early infancy. In 2016, when I was carrying out a research on pregnancy apps, I analyzed 3,570 reviewers' comments of ten of the most downloaded apps in the UK and the US in 2015. The analysis enabled me to shed some light on parents' attachment to the data they produce through data-tracking apps. I remember one example in particular. On March 14–15, 2016, one of the apps I studied updated its features, and many users lost their data. The app was inundated with comments. Users mentioned their outrage and requested to have their data back. I remember being fascinated especially by one user who wrote how upset she was about losing her "kick count data" and ended her comment with an angry, "Shame on you!"

WHY DATA TRACKING MATTERS?

Some would describe parents' attachment to data as a form of "data obsession" or "data fetishism" (Speed 2019). Yet anthropological literature on the fetish shows that humans often don't fetishize objects (or data) as a form of lunacy, they fetishize them because these objects embed, represent, and remind them of their social relationships (Graeber 2007). At times parents form a deep emotional bond with their data-tracking technologies because these technologies enable them to live and experience the important relationships in their life. A user of one of the pregnancy apps that I analyzed, for instance, described the app as her "best buddy" helping her through "all the stages of pregnancy."

Data tracking for family life matters, and it matters for a variety of reasons that reflect the plurality of data that we produce. We record data because we want to capture instances of our experience, and we feel an emotional bond to some of the data that we produce. On the day I discovered I was pregnant with my first daughter, for instance, I took a picture of my body with the Photo Boot app on my MacBook. Every week for nine months I documented my pregnancy. I also used the computer to take screen grabs of my sister's and my friends' reactions to the news of my pregnancy after I told them on Skype. I saved all my pictures in a folder, titled "family." That data was so special to me—irreplaceable. When I thought I had lost it because my computer had crashed, I felt lost, angry, and terribly upset; when I was told by Craig, my tech-savvy friend, that no harm had been done, I was excited and relieved.

When we think about family life, therefore, we cannot ignore the emotional dimension and importance of data. This dimension is clear if we think about our attachment to photos. Photos capture life moments; they are evocative of specific atmospheres and patterns of our lives. They tell our story and the stories of our children; they tell us about change and the passage of time. This is very true also for digital photographs. The shared assumption for a while has been that digital photography and online photo sharing have replaced the traditional role of photographs in family life as media of preservation and memory. MacDonald (2018) has shown, however, that digital photography and photo sharing can have a profound meaning for families, and this meaning relates to the construction of a collective memory.

It is not only photos that matter. The details of everyday life that can be recorded and captured, even the most mundane and ordinary details, also count. This is why data tracking can be very meaningful for families. The practice of documenting mundane details of everyday life has always existed before the advent of social media and other data-tracking technologies, especially among families (Humphreys 2018). By looking at the social role played by diaries in the eighteenth century and throughout the nineteenth century, Humphreys (2018), for instance, shows that everyday life was documented with precision on personal diaries. These diaries were interactive because they were often shared with family and friends and were also mobile, because diarists often used pocket diaries to record life in real time.

Data tracking in family life has a long history and a profound emotional dimension. In the past, parents who used journals to track their families' routines owned and controlled the data that they produced, because like my mother they owned their journals and often kept them in safe places. Today this data is stored, processed, and profiled in ways that escape parental knowledge and control.

In March 2019, for instance, the *British Medical Journal* published an international research, which demonstrated that out of twenty-four mHealth apps, nineteen shared user data with parent companies and service providers (third parties). They also showed that third parties shared user data with 216 fourth parties, including multinational technology companies, digital advertising companies, telecommunications corporations, and a consumer credit reporting agency. Of 216 organizations, only three belonged to the health sector. The paper also demonstrated that the data was shared with different Big Tech companies, including Alphabet (Google), Facebook, and Oracle, who occupy central positions within data-brokering networks because they have the means to aggregate and re-identify user data (Grundy et al. 2019).

As parents buy into the promises of data tracking, they produce large amounts of children's data that is then archived, analyzed, and sold; hence, they play an active role in the datafication of children. However, the datafication of children does not only occur because parents play an active role and use data-tracking technologies. In fact, not all parents use pregnancy and baby apps, and many parents that I met during my research often complained that data tracking was "too much work."

Yet as the next part will show—even among those families who do not use mHealth apps or wearables—children are nevertheless datafied from before birth. This is because they are exposed to the business models and the data-brokering practices of surveillance capitalism, which enable companies to track children from before conception.

THE DATAFICATION OF THE CHILD: NO WAY OUT

I never used an app, but Google and BabyCenter—and all their third and fourth parties—knew I was pregnant with A before my family knew. I found out I was three weeks pregnant on holiday. Google had to answer my following concerns: "Can I use a hot tub in early pregnancy?"; "Abdominal cramps in early pregnancy"; "Miscarriage risk when flying." I found most of the answers I needed on BabyCenter, and because I had already studied their data policy (Barassi 2017a), I knew that my data was recorded, processed, and shared with third parties. But what could I do? I needed answers, and online searches are part of my life. Yet it made me feel awkward to know that targeted advertisers knew I was pregnant and started profiling me before I even had the chance to share the news with my mom, dad, and sister.

Under surveillance capitalism, the profiling of the pregnant woman and the datafication of the unborn has become unavoidable. In 2014, for instance, Janet Vertesi, a sociology professor at Princeton, did an experiment and tried to see if she could keep her pregnancy secret from the bots, trackers, cookies, and other data sniffers online, which feed the databases that are then used for targeted advertising. She was aware that pregnant women are tracked more than other users, because data companies believe that to be able to identify one pregnant woman is worth as much as knowing the age, sex, and location of up to 200 people.

In her article published in *Time* magazine in 2014, she explained that trying to hide her pregnancy made her look and feel like a criminal, because she had to employ different tactics such as using Tor as browser to access content of the BabyCenter. She thus came to the conclusion that trying to avoid becoming a "pregnant data subject" made her look not only like a rude family member or an inconsiderate friend, but like a bad citizen (Vertesi 2014). For mothers-to-be like Vertesi, it is impossible not to be tracked and profiled as data subjects. In fact, the tracking of pregnancy and early infancy has become a fact of our datafied lives under surveillance capitalism, which families cannot avoid.

Early infancy is just the beginning of the datafication of children and their families. As children grow up, parents search Google with age-specific queries and land on web pages such as those of BabyCenter.com or Baby-Centre.co.uk that are already structured around age (getting pregnant, baby, toddler, preschooler, big kid) and share this information with others.

The BabyCenter LLC privacy policy explicitly states the data collected is shared with their affiliates and third-party partners. It also mentions that the company has "A direct relationship with the following ad technology partners who may have information about you that may be used to personalize the advertisements you receive when using our Service: Google; Amazon; AppNexus; Brightcom; District M; DoubleVerify; Index Exchange; Live-Intent; OpenX Technologies; Salesforce; Sizmek; Smaato; Sovrn; Teads.TV; YieldMo; bRealTime/EMX" (BabyCenter LLC 2019, para. 4). Hence critical questions emerge regarding whether data companies know and share the exact age of children and have the technologies to track and follow them as they grow up.

Children are being datafied in ways that were not possible before because families use data-tracking apps and wearables, they search for information online, and post photos on social media. However, the datafication of families and children is not only happening because families use apps, search engines, or social media, but also because, with the extension of surveillance capitalism, the society around them (e.g., the school, doctor, bank) is increasingly becoming automated and data driven. From doctor's appointments to schools, from supermarkets to home technologies, family life is being surveilled, tracked, and analyzed in almost unimaginable ways.

THE DATAFIED FAMILY AND SURVEILLANCE CAPITALISM

One of the big changes brought by *surveillance capitalism* is the introduction of the cultural belief that data offers us a deeper form of knowledge. According to Mayer-Schönberger and Cuckier,

> Data was no longer regarded as static or stale, whose usefulness was finished once the purpose for which it was collected, was achieved. … Rather data became a raw material of business, a vital economic input, used to create a new form of economic value. In fact, with the right mind-set, data can be cleverly reused to become a fountain of innovation and new services. (Mayer-Schönberger and Cuckier 2013, 5)

This is why we have seen a radical transformation in society. Governing institutions, educational bodies, health care providers, businesses of all kinds and multiple other agents have started to turn every aspect of everyday life into data (Mayer-Schönberger and Cuckier 2013). In other words, they have bought into the logic of surveillance capitalism and have started to gather enormous quantities of personal data. During the Child | Data | Citizen project, I realized that parents were aware of this transformation. When I interviewed Mike in 2017, I asked him if he could imagine the data flows that came out of his family life. Mike, as mentioned in the previous chapter, was married to Zoe, had two children aged 12 and 5 years, and lived in Los Angeles. When I asked him the question, Mike laughed, looked up, and said, "massive amounts; unimaginable amounts," and then added

> M: If one thinks about the data we produce in my family, we are talking about all types of data: location and time stamp, things that I like on social media or that I share, our browsing history, our health, online banking, home bills (I have automated everything that I possibly can), school data, all our entertainment choices, everything we buy. We don't have home technologies or at least not yet and no Amazon Alexa, but I think it's coming.

Mike was aware that his family was being datafied not only because he used specific technologies (e.g., social media), but also because all the services that he encountered in his life were increasingly becoming data driven and automated (e.g., the energy suppliers, schools, doctors, among others). When Mike described the different data flows that came out of his family, he told me that he could remember a time when "There was not so much data out there." In the last 10 years, he believed, something changed. There was a shift in the ways in which "society understood and valued personal data" and consequently in the amounts of data that was produced and collected. He described this transformation as a gradual push by companies, firms, and institutions to make you produce more and more data so that they could track it and profile it.

Few months later I was sitting in Dan's living room in London. Dan was the father of two children, aged 7 and 5 years. Dan had left his job in information technology in advertising to become a stay-at-home dad and support Jill's career in marketing. That day when I asked Dan how he understood the datafication of his family, I was immediately fascinated that he used almost the same words as Mike did. Like Mike, Dan talked about a massive change: "Everything, just everything, every shop, every bank, every organization pushed you to download an app or register."

It was striking that both of them perceived this transformation as gradual, as something that happened over the last 10 years and they gradually adapted to. Dan followed the transformation because he was working in the advertising business, and for this, he felt that he "was ahead of the game" in realizing that massive amounts of personal data (and family data) were being produced, collected, and processed. Mike realized that "something was going on with data" more than five years ago when he came to the conclusion that everything online was easily tracked and that all services were becoming digitized and data driven. For him the datafication of family life was something that he saw coming.

Mike and Dan perceived the transformation as gradual. Yet there is something profoundly unequal about the different ways in which the datafication of society has impacted highly educated or high-income families on the one hand and low-income or less educated families on the other.

THE UNEQUAL IMPACT OF THE DATAFICATION OF FAMILIES

Both Mike and Dan were highly educated with high income and experienced the datafication of family life as a gradual transformation. Their experience of change radically clashed with the experience of those parents who came from a low-income or less educated background and who told me that they experienced the transformation as sudden, unexpected, and difficult to deal with.

Alexandra, for instance, was a low-income Hungarian immigrant working in London. Alexandra was married to Sid, who was from Nigeria, and had two children aged 8 and 10 years. She and her husband suddenly realized that their data was constantly collected because all the services around her were being digitized.

A: [There is so much data] because everything changed, everything went online. Like online banking but I am resisting it, but they make it impossible to go to the branch because they are closing branches down, so I end up using it. I don't do much with it because I don't understand it much.

V: Did you perceive the change elsewhere apart from banks?

A: Yes everywhere, you know in the shops the award cards. I have few of them. But also, the kids' school now everything comes online by

> email, on a daily basis. It's useful because there is so much information in relation to the school, so it is easier for me to be informed. But now they are trying to set up an online payment system so that I can pay the school meals, the school trips, etc. But I don't use it because I don't have online banking and so I prefer to pay every month. It's that it is too difficult for me. I am sure that if someone would sit down with me and explain it to me, then I would get it. At the doctor I use the online booking; again I don't really know how to use it, and it is frustrating. But our GP [general practitioner, doctor] surgery has become so busy that sometimes I need to wait 4 or 5 weeks to see a doctor, but if I go online, I would find an appointment sooner.

For Mariana, like Alexandra, the change was quite sudden, and she felt that she lacked the skills to cope with it. Mariana was a Mexican immigrant who worked as a cleaner in Los Angeles and was a widow with four children, aged 11 to 23 years. "I know nothing about technologies" she told me when I started her interview. In contrast to her children who "were always on their phones or tablets," Mariana managed not to use an email or a smartphone up until 2017. Then something changed; she was forced to go online. The school of her youngest children (11 and 13) started to rely on an online platform for homework and internal communications. She felt that she had no choice; she had to learn because she wanted to support them. Once she started going online, she suddenly became aware of social media and of how much information her children were posting online. "It was shocking and worrying, I did not know what to do."

Mariana's and Alexandra's interviews were thus strikingly different from Mike and Dan. For them, the datafication of family life—which went hand-in-hand with the digitization of services, was more of a shock than a gradual process—they felt isolated and felt that they lacked the knowledge and skills.

The different experiences of Mike, Dan, Mariana, and Alexandra speak to the fact that social inequality plays a fundamental role in the way in which the datafication of family life and children is experienced and dealt with. Although Mike and Dan felt that they were ahead of the game or saw it coming, Mariana and Alexandra perceived the change as shocking and sudden and as something that was forced on them.

This inequality was also reflected in the ways in which they understood data privacy. If we compare Mariana's and Mike's understanding of the implications of the datafication of family life, they are strikingly different. Mike, like so many other parents that I worked with, felt that "he would prefer for the data to be private" but that "he had nothing to hide." For Mariana, "the data out there was scary," and a real concern, as she could clearly see how it could impact and harm her and her family.

I will discuss the privacy divide at length in chapter 9 and will engage with critical questions about algorithmic bias and systemic inequality. Here, however I want to focus on the fact that it is precisely because the society around them is becoming more and more data driven that parents no longer have a choice *but* to sign terms and conditions and give their consent for the lawful processing of children's data. Despite that current data protection regulations focus a lot on parental consent, under surveillance capitalism, the notion of informed consent is exceptionally problematic. This is because, in the everyday life of families, digital participation is no longer only voluntary but increasingly more *coerced* (Barassi 2019) as parents are forced to comply with data-driven and automated systems.

DATAFIED CHILDREN AND THE PROBLEM OF CONSENT

One day in Los Angeles in 2017, my family was invited to join a group of friends at an indoor play area in a shopping mall in the Valley. That day we had been stuck in traffic for more than an hour. When we arrived at the play area, P (who at the time was 4 years old) and A (at the time 6 months old) were hungry, frustrated, and whiney. As soon as I reached the counter to purchase the tickets, the employee behind the desk asked me to write down my children's names, dates of birth, home address, my phone number, and email. I was also given the option to note my social media accounts. I felt annoyed. Why did I need to provide all that information simply to be allowed access to a play area? I really did not want to write down my children's birth data and home address.

I asked the employee at the counter why they needed all that information. The employee looked baffled. Behind me there was a long line of parents and screaming children eager to get in. She looked up from her screen and replied: "It's for insurance purposes." I felt uncomfortable, and I asked whether I could opt out. She looked up again, this time confused, and told me: "The only way you could opt out is not to purchase the ticket." I looked at P, who had just taken her shoes off and was waving at her friends. I felt I had no choice and gave away all our personal details without even reading the terms and conditions. That evening, when I got home, I found myself writing in my fieldnotes about that experience, about how it affected me and made me feel. The entry is one of the first entries in my database that explore parents' relationship with terms and conditions.

On an average family day, parents join new services, download new apps, connect with others on social media, or buy the latest home devices. In doing this, they sign off on the terms and conditions of a variety of services and give their meaningful consent that provides companies the right to lawfully process their children's data. During my research, however, I came to the conclusion that—as it happened to me that day at the play area—the consent that most parents give is not informed or meaningful.

A great majority of the parents that I interviewed did not read the terms and conditions. This is not surprising. To read data policies requires an enormous amount of time that parents often do not have. In 2008, McDonald and Cranor (2008) calculated that reading all the privacy policies of the websites that users encountered in daily life would take approximately 201 hours a year for each US user. Their calculation was based on the fact that to read an average policy takes from 8 to 10 minutes and that at the time, according to the data of Nielsen/Net Rating, an average US user would visit 119 websites *annually*. Our digital environments have changed dramatically in the last 10 years, and there is no way to calculate how much time it would take for an average user to read all the privacy policies.

I had a vivid example of this during my research. One evening, in September 2018, I was in London and just got back from a night out with friends. I sat down at the computer to write a couple of emails. I checked Facebook and saw a link to an article on Asia Argento and the #MeToo Movement. I clicked on the link that brought me to an Italian magazine website. To comply with General Data Protection Regulations of the EU, the website asked me if I wanted to opt out personalization and find out more about the third parties with whom they were going to share my data. I clicked yes, and there was a long list of companies across the globe that had access to that insignificant click. Next to the name of each company there was a link to its privacy policy. I counted how many privacy policies I had to read to find out what happened to my personal data for a simple web search, and the result left me speechless: it was 439.

Parents do not read terms and conditions because they would never find the time to read them. They also do not read them because they feel they have no choice: either they agree to the terms of service or they would not have access to important services in their life. This lack of choice is understood as the *privacy trade-off* in which users give up personal data just to be able to access specific platforms and services (Turow et al. 2015; Hargittai and Marwick 2016). Draper and Turow described this act as *digital resignation* because people resign to give up their personal data to enjoy a service. They argue that digital resignation has not only become a shared and normalized practice among users, but it is also constantly cultivated by corporations who encourage and reinforce it (Draper and Turow 2019).

The day that I found myself signing off the terms and conditions of the play area, I resigned to give up the data of my children. I felt the pressure: either I agreed to give up that data, or I had to tell my daughter that we were not going to meet her friends. In my daily life I am constantly coaxed into acts of digital resignation. Although I try to protect my children's privacy, on a daily basis I buy into the privacy trade-off. During my research, however, I realized that surveillance capitalism does not only rely on the cultivation of digital resignation but also on the *systematic coercion of digital participation* (Barassi 2019). This is because, in many instances, the parents I was working with, and myself included, were not just resigning to digitally participate, they were actually forced to do so.

THE COERCION OF DIGITAL PARTICIPATION

In 2018, during my time of fieldwork I was traveling through London Heathrow airport with my daughters. As soon as we reached security, an officer looked at our passports, asked P (at the time 4 years old) to look up and not to smile for the camera. After she dutifully obeyed, he took a small, black device and scanned the face of A (at the time 9 months old). I felt uncomfortable and violated. I asked him why he was scanning my baby's face. He looked annoyed and explained that it was for security purposes. Facial recognition was being used as an anti-terror measure to prevent passengers traveling on international flights swapping their boarding cards with passengers from domestic flights. Obviously, this is not a new security measure. Facial recognition technologies at Heathrow were introduced more than seven years earlier. Yet, I felt that on that day the pervasiveness of facial recognition technologies at Heathrow was more evident than other years, because of the introduction of the technology by British Airways.

As I walked through security, I felt a chilling feeling in my bones and wondered whether I would have felt worse if the officer took my children's fingerprints. Maybe I would have been more outraged. Yet the result is very similar. Like fingerprints, the face, iris, or even the ear canal are all used now to carry out ID checks; they are all classified as biometric data. What also bothered me was that nobody informed me about the ways in which that data was going to be used. I was forced to comply, and I could not opt out.

One of the main changes brought about by surveillance capitalism is that the institutions and businesses that individuals encounter in their everyday life (e.g., health providers, education institutions, local governments, police and border agents, and so forth) are increasingly relying on data-driven services. In this context, parents are *coerced* into digitally participating.

The term *coercion* is used here with reference to classical understanding of voluntary and coerced citizen participation, in which coercion does not imply the use of force or violence, but refers to forced compliance with structures, policies, and regulations (Milakovich 2012, 31). Parents are forced to comply and provide their data, otherwise they risk physical consequences (e.g., let's imagine if I refused to have my baby's face scan taken at the border) or social and personal consequences in the sense that they would be left out from important areas of social life (e.g., refusing to use Google Classroom that has been introduced in their children's school during the COVID-19 crisis).

When we reflect on these examples of families dealing with a datafied society, we realize that we need to problematize the very idea of parental consent that is key to current data privacy regulations for three main reasons:

1. Data regulations focus on informed consent and the importance for policies to be clear and transparent (e.g., GDPR). Yet even if privacy policies and terms and conditions become transparent and user friendly, parents cannot possibly read all the policies. They often have no choice but to sign them, so their consent is not meaningful or informed.
2. Companies are constantly trying to gather more data by manipulating users into giving up their consent. A key example is represented by the 2018 investigation by the Norwegian Consumer Council (NCC), which has shown that Facebook, Google, and to some extent Microsoft, were running consumers out of privacy-friendly options on their services in an unethical way.

3. Public debate often focuses on the personal data that is willingly disclosed by users. Yet little attention is being given to—as we have seen with reference to the tracking of children—companies and other agents that are relying less on data we provide and instead are looking at data they can observe, derive, and infer (Privacy International 2018). Today consent has become a problematic issue, one that needs to be understood by looking at our everyday realities and by critically reflecting on our everyday practices.

As I will argue in more detail in chapter 8 current privacy regulations are failing children and families. They focus on ideas of transparency and choice, but we no longer have a choice; our data is constantly taken from us and our children and is used to make decisions about us. In the age of artificial intelligence (AI) and data-driven systems, the issue at stake is no longer only about individual privacy and data collection, but it is about the type of assumptions and conclusions that are made about individuals on the basis of their data traces. We need to question the social and political implications of building a society where data traces are made to talk for and about citizens across a lifetime. What happens when the algorithms get it wrong or the narratives constructed about individuals are biased, inaccurate, and discriminate against them?

CONCLUSION

In the age of surveillance capitalism families and children are being datafied from the moment of conception. This is not only because parents rely on new mHealth technologies or search engines, which sell their data to advertisers, but also because the society in which they live has become data driven and automated. The datafication of family life has become a new and unavoidable reality for many. Yet there is something profoundly unequal about the different ways in which this datafication impacts high-income or highly educated families on the one hand, and low-income or less educated families on the other.

Children today are the very first generation of citizens to be datafied from before birth, and we cannot foresee—as yet—the social and political consequences of this historical transformation. As the next chapters will show, what is particularly worrying about this process of datafication of children is that companies like Google, Facebook, and Amazon are harnessing and collecting multiple typologies of children's data and have the potential to store a plurality of data traces under unique ID profiles. It is for this reason that we need to break down children's different data flows and analyze the practices, beliefs, and structures that make these flows possible. Only by doing so can we grasp the complexity and breadth of the datafication of children. In the next part of the book, I will be focusing on four different data flows: health, education, home life, and social media.

3

HEALTH DATA: TRACKING THE HEALTH OF FAMILIES

On July 21, 2017, I slipped and fell on the bathroom floor in my home in Los Angeles. I was thirty-eight weeks pregnant. After calling my doctor's office, I was rushed to hospital. The fall had induced my labor, yet contractions were mild and, as we waited for my doctor to arrive, I chatted with the nurses. Both the nurses had kids, and we talked about my previous labor and other family-related issues. As I was chatting to the nurses, a young assistant came into the room with a file in her hands and a smile on her face. She had been informed that I had agreed to donate the cord cells to the hospital and handed me the forms to sign.

I started reading the terms and conditions and noticed a little box that I needed to tick, which read something along the lines of, "I agree for the cord cells to be used for any future research." I looked up and asked her whether there were more specific terms that I could agree to (e.g., I agree for the cells to be used for curing patients, or something similar). I did not feel comfortable that the hospital could use the cells for "any future research." I joked and said:

V: As far as I know, the hospital could clone my baby if I agreed to such nonspecific terms.

She was baffled and embarrassed that she couldn't answer my questions.

A: It's just a form, you just sign it.

I did not have the mental space or time to discuss the issue further. My contractions were intensifying, so I decided not to donate the cord cells. The nurses were intrigued and kept saying, "You do have a point, you really do."

I started this chapter with this very personal anecdote because it was during my experiences of pregnancy and early motherhood that I realized not only how much health data was collected from me and my children, but also how natural and accepted it has become to sign off on terms and conditions that give away important health details and genetic data.

Historically, the health care industry has always collected large amounts of personal data to advance medical research or personalize care. Data in the health care industry has always been used to achieve important social and medical goals and to improve care for society as a whole. For this reason, patients have historically lacked any control or say over their heath data; they never owned their data. The data of the unborn and prospective mothers is a crucial example of this. This data has a social life of its own, where parents (and children) have little control or say (Lupton 2013).

Although the collection of health data has always existed, the advent of big data, AI, and surveillance capitalism has radically transformed the production, sharing, and processing of health data. Over the course of the last decades, in both the UK and the US, governments and the private sector have invested large sums into the datafication of the health sector. At the same time, families have started to use a variety of mHealth and data technologies, which are designed to harness enormous quantities of health data from family life. The production and harnessing of health data can have a key impact on children's lives, and we need to critically reflect on the multiple ways in which the Big Tech companies are collecting and using this data. These questions are particularly important at this historic time when tech giants are playing a key role in health-tracking practices that are becoming more diffused due to COVID-19.

BIG DATA IN HEALTH

After A was born, I started suffering from recurrent headaches and visual problems. I first went to my family doctor in Los Angeles, who diagnosed me as having the symptoms of chronic migraines. He thus prescribed me a migraine medicine, which I had to take regularly as soon as I felt that the migraine was starting. I tried the medicine for a week, but it nauseated me and I couldn't perceive any difference in the recurrence of my headaches. So I stopped taking it and forgot about it. Six months later, a family friend who is a doctor told me to go and see my OB/GYN doctor and check whether my alleged migraines were caused by a hormonal unbalance. I followed his advice, and my OB/GYN put me on a pill. Since then I have not been suffering from recurrent headaches or visual problems.

When I went for a follow-up with my OB/GYN, I was enthusiastic about the disappearance of my recurrent headaches. Sitting on a little stool with his laptop on his knees, my doctor asked me whether I was still talking my migraine medicine. I was surprised and asked

V: What medicine?
D: Your migraine medicine, the one that Dr. XXX prescribed you.
V: Oh, I forgot about that. But you can *see* that?
D: Of course, everything is here in your health record.

I immediately worried that the data about my migraines would impact my insurance premium, and then added

V: You can really see that? I had forgotten about it. But this worries me. It's like the big brother of health, and they could profile me with a chronic condition.
D: I know, they have everything, and it's scary.

My doctor went on to tell me the story of his wife who had problems with the health insurance at one point just because "She was profiled as suffering of acne." He then added

D: This is why I never agreed to give my genetic data away; my DNA is the only part of my body that is not out there yet.

My doctor made that comment because he knew what I was doing for a living and about my research. Yet he also made that comment because he had firsthand experience of electronic health records (EHRs) and health data processing, and he also understood what had changed over the last decade or so.

Historically, the health care industry has always produced large amounts of health data, not only data that was aggregated anonymously for research purposes but also data that involved the personal history of a single patient and his/her family. Translation of ancient Egyptian hieroglyphic inscriptions and papyruses indicate that the use of medical records has been a common practice since ancient civilizations. Yet we needed to wait until 1900–1920 for the first paper medical records to be steadily introduced (Evans 2016) and then for the 1960s to see the first electronic medical records. The first EHRs were noncentralized computer systems such as the Health Evaluation through Logical Programming (HELP) launched in 1967 by the Latter Day Saints Hospitals in Utah, the Computer Stored Ambulatory Record (COSTAR) developed by Harvard Medical School, and the Medical Record developed by Duke University in 1969.

Even if historically medical records have long existed (including EHRs), the last two decades have witnessed a radical digitization and datafication of the health sector in both the UK and the US. In the US at the beginning of the 2000s, the digitization of health has started to become a national issue crossing businesses and state boundaries. In 2004, President George W. Bush launched an initiative for the "widespread adoption of Electronic Health Records."

In 2009, The Health Information Technology for Economic and Clinical Health (HITECH) Act was introduced to incentivize the adoption of EHRs at the national level. At the time, former president Barack Obama addressed the national importance of EHRs by maintaining that they would enable the health care industry to cut waste and prevent medical errors (Rahman and Reddy 2015). All these government incentives enabled the creation of more centralized and standardized health data infrastructures for the analysis, processing, and sharing of health data. They also enabled the creation of individual health records that could gather large amounts of an individual's health data from a plurality of different sources.

Today, according to the US government's Health IT Portal, an electronic health record—like the one that my doctor accessed—contains the following information: a patient's medical history, diagnoses, medications, immunization dates, allergies, radiology images, and laboratory and test results. It can bring together information from current and past doctors, emergency facilities, school and workplace clinics, pharmacies, laboratories, and medical imagining facilities (HealthIT.gov n.d.). In other words, EHRs can contain the health life of individuals from conception to death, which is collected from a variety of different sources.

Also in the UK, over the same historical period, the health sector became datafied. Although the first steps into the digitization of health and the creation of EHRs date back to the 1970s—and John Preece was allegedly the first general practitioner to use a computer in his practice (Wainwright 2006)—the 2000s were key to the datafication of the UK's health system. In 2002, the UK Department of Health allocated £18 billion for the development of the National Health Service (NHS) technology (Chantler et al. 2006).

In 2013, the Health and Social Care Information Centre (HSCIC; usually known for its branded name, NHS Digital) was established as "an executive, non-departmental public body and the national provider of information, data and IT systems for health and social care." The center was established as a direct result of the 2012 Health and Social Care Act with the goal to improve health and social care in England by putting technology, data, and information to work. Their strategy and mission since the beginning was to "create a new architecture" (or in other words, a data infrastructure) that enables standardized practices of data gathering and data analysis throughout the health sector at national level (HSCIC 2015).

In both the UK and the US, therefore, over the last decade or so, we witnessed the development of governmental strategies and incentives for the datafication and digitization of the health sector that have left little choice to families but to comply. As shown in chapter 2, during my research I met different parents who felt that they had no choice but to adapt to online portals. Alexandra, who lived in a low-income neighborhood in south London with her two children and her husband told me that she struggled with online systems, but she also mentioned that she was forced to find her way around the online booking system of her general practice (GP) surgery. Whenever she called to book an appointment, she would only get an appointment four or five weeks later, but when she booked it online she would get one straight away. Other parents that I worked with in the US shared similar experiences. These experiences bring us back to the very idea of coerced digital participation (discussed in chapter 2) and remind us that families are forced to comply with the datafication of health.

THE DATAFICATION OF HEALTH AND WHY DATA PRIVACY MATTERS

During the research, I met many parents who, like me, were gradually adapting to the datafication of the health sector and signing off terms and conditions that gave away their and their children's health data privacy. Some of the parents—especially those who came from a well-educated background—praised the datafication of the health sector as beneficial, because it made life much easier. Nicole, who lived in Los Angeles and who I mentioned in the previous chapters, was often in and out of hospitals because both she and her daughter suffered from chronic health conditions. During the interview, she told me that she loved online portals.

S: I love them. I only just started using them. When I was going to the hospital before, I never signed up for the online portal, but now that I am doing stuff at a different hospital, everything is online.

V: What do you love about it?

S: Oh I just love it, I can see test results, I can email my doctors, I don't have to wait for them to call me. There is more transparency.

Like Nicole, Caty, who lived in North London with her 5-year-old son, described the change as a positive change. Caty suffered from asthma, and she was thrilled because she could ask for her prescription online rather than having to go to her health practice.

Both Caty and Nicole were clearly benefiting from the use of the digital health portals provided by their health care providers, yet they both questioned the privacy implications of the health data they shared. Nicole was worried about insurance profiling. Accustomed as she was to have to negotiate constantly with her health insurance, she worried (like I did) that her health data could be inaccurate and could be used to profile her in ways that would impact her insurance premium. Caty, instead, had a completely different understanding of health data privacy. For her, the risks of health data were not to be found in individual profiling and insurance claims, but rather in the ways in which data-brokering practices would impact the democratic mission of the National Health Service.

C: I am happy for them to use my data for health research. But then a few years back there was a change and they said that GPs could now collect a lot, a lot, of data and share it. I think it was the care.data program. I wrote to my GP and opted out, because I was uncomfortable with the fact that the central government would have access to all that data. I had issues about how that data was kept and the security of it. But also I had wider social concerns about how data is used to inform policy development and funding. You know they draw pictures and conclusions that don't actually reflect the actual reality.

What Caty was mentioning was the Care.data program, which was launched by the HSCIC in 2013 and was meant to extract the data from GP surgeries for the creation of a central database. Members living in England, like Caty, could choose to opt out by writing to their GP surgery. Since its launch, the program became highly controversial. Privacy groups and other governmental and nongovernmental agents in the UK argued that the program impacted patient privacy because the NHS could not guarantee that the data would not be shared with insurers or third-party organizations. In 2016, after years of controversy, the program was scrapped.

Although in different ways, both Caty and Nicole were concerned about the data harms that could emerge from the sharing of health data. As the health sector becomes more and more datafied, we need to be aware of the plurality of data harms that can emerge from the sharing and processing of health data. In 2017, for instance, *The Guardian* newspaper in the UK revealed that the Home Office used NHS records to track down immigration offenders (Travis 2017). In November 2018, following legal action from the Migrants' Rights Network, the UK government agreed to suspend most of the data-sharing arrangement with the NHS and limit its use to tracing those being considered for deportation because they had committed a serious crime.

Parents often share their health data and the health data of their children with their doctors because they trust that their health data is protected and used only for medical purposes. Yet often they do not consider the implications of data-sharing agreements in the age of surveillance capitalism. What they also do not consider is that in the UK and the US—despite that the two health systems radically differ in their approach to public/private health provision—their strategies for datafication and digitalization shared a great similarity. In fact in both countries, the datafication of the health sector is made possible through the reliance on privately owned software and technologies and the opening up of the sector to Big Tech.

IS BIG TECH TAKING OVER THE HEALTH SECTOR?

In both the UK and the US, privately owned companies have started to gain access to large amounts of health data that is gathered through the datafication of the health sector. In 2014, Kaiser Permanente, the California-based health network that had 9 million members, was believed to have between 26.5 and 44 petabytes of potentially rich data from EHRs, including images and annotations (Raghupathi and Raghupathi 2014). In 2018, The EMIS Group in England, whose health suite is used by 56 percent of GP organizations, held more than 40 million records. In November 2018, different health-related professional websites reported the news that the EMIS Group was going to migrate the data of its 40 million users to the Amazon Cloud (Heather 2018).

As the health sector becomes more datafied, and as more private companies are gaining accesses to health data, critical and ethical questions emerge on the ways in which this data is collected, processed, used, and shared. In particular, we need to critically explore and consider the new role played by Big Tech and the implications for data privacy.

The example of Alphabet/Google is perhaps one of the most interesting. In 2015 Google restructured into Alphabet, and with the restructuring of the company it was clear that one of its strategic priorities was precisely to invest in health care. According to a research report by CB Insights (2018), Alphabet relies on four different subsidiaries to pursue its health care mission: Verily, DeepMind, Calico, and GV.

Verily is focused on using data to improve health care via analytics tools, interventions, research, and more (e.g., partnering with existing health care institutions to find areas to apply AI, especially via its Study Watch, a wearable device that captures biometric data). DeepMind is based in London and works closely with the UK National Health Service; its mission is broad in the sense that it aims to foster artificial intelligence research in health care. Calico focuses on combating aging and age-related diseases. GV is the health care venture arm of Alphabet and invests across different sectors.

So far Alphabet's health care mission has been focused on tackling specific diseases (in particular, eye disease, diabetes, heart diseases, Parkinson's disease, and multiple sclerosis) through the development of the following methods and technologies:

1. Data generation: Using AI to tackle specific diseases through data generation (e.g., gathering large quantities of health data including the data produced through wearables, medical imaging, and other sources);
2. Disease detection: Using AI to detect patterns and anomalies in heath data sets that signal some disease; and
3. Lifestyle management: Supporting patients with specific diseases to make lifestyle changes.

Alphabet has been investing also in "powering the health care data infrastructure layer." In other words, the company is playing a fundamental role in the development of the standardized and centralized data infrastructures mentioned previously. A key example can be found in the role played by DeepMind for the NHS in the UK. The company is based in London and works closely with the National Health Service to build a new data infrastructure that brings together the information of different EHRs with the data gathered through hospital equipment and doctor's notes.

The example of DeepMind in the UK is very interesting as it exposes some of the questions and implications that arise when Big Tech enters the health sector. In 2015, DeepMind and the Royal Free NHS Foundation Trust made a data-sharing deal that implied that the Google company had access to the sensitive health information of 1.6 million patients. In 2016, the Information Commissioner's Office (ICO) launched a probe on the deal and found out that the agreement failed to comply with data protection laws and regulations (Kharpal 2017).

Yet the controversy around DeepMind and data privacy did not stop in 2016. In November 2018, the company announced that DeepMind would be absorbed into Google Health together with other areas working in health care such as the Google Brain, Google Fit, and Nest, which focuses on harnessing health data from the home. This announcement reversed an earlier pledge that DeepMind would not be absorbed within the company. Different privacy groups in the UK highlighted the risks of such move by arguing that there was no way to assure that in the future the company would not integrate health data into unique ID profiles and the other data gathered through services such as Google Chrome, Gmail, Google Docs, Google Maps or YouTube (Kahn and Lauerman 2018).

Of course, Google/Alphabet is not the only Big Tech company investing in the health sector. Apple, Amazon, and Facebook have long been investing in health care. At times these investments face backlash. In 2018, for instance, Facebook pursued a data-sharing agreement with eight top hospitals in the US. The sharing agreement failed because of privacy concerns at a historical time in which Facebook was in the midst of the Cambridge Analytica Scandal (Farr 2018). Yet such a backlash was an exception, and companies like Apple, Amazon, and Facebook are making more and more deals to tap into the datafied health care sector.

As Big Tech companies are trying to play a fundamental role in the datafication of the health sector, we must realize that they have the potential to gain competitive advantage over others. Through their technologies, services, and data-brokering deals, these companies are already gathering unimaginable amounts of health data from family life, including the data of children. What is particularly concerning is not only that they have the technologies and business models that enable them to gather a lifetime of health data (from conception to death) but they also have the potential to bring all the health data together with other personal data into unique ID profiles.

These companies are already trying to harness two different health data flows in family life: web searches and mHealth apps.

FAMILY LIFE AND DR. GOOGLE

I have searched Google countless times for health information about my children. Anytime one of my children would get hurt or had a bad temperature I would go to Google to find out what to do. Should I call a doctor? Should I take them to urgent care? What are the red flag symptoms? I also searched Google to find answers to any developmental or psychological concern I had, and I used key words such as "twelve-month-old not walking" or "anxiety symptoms in toddlers." I searched Google and a plurality of other websites from BabyCenter to the NHS website to soothe my anxieties, and I never really gave it a second thought. When I started this project, I realized what happens to my health data and the health data of my children, and my attitude to health-related web searches has radically changed.

Parents, like me, regularly turn to Google or other search engines, for health-related questions about their family and their children. Yet all this data is constantly collected, processed, and shared in ways that families are often completely unaware of or rarely think about. When parents search Google for specific symptoms, they navigate a variety of websites that collect data and send that data back to Big Tech companies as well as to an enormous number of other companies. Tim Libert, a researcher from the University of Pennsylvania, explained what happened when he did a web search for HIV and landed on the page of the Centers for Disease Control and Prevention (CDC). He found out that from the HIV page of the CDC website, third-party requests were made to the servers of Facebook, Pinterest, Twitter, and Google (Libert 2015).

The fact that a public sector website such as the CDC, which was the first website that parents usually consulted in the US, would share such sensitive data with private companies is really unsurprising in the current data environments. In March 2019, researchers from Cookiebot—a company that detects trackers on websites—joined forces with campaigners at the European Digital Rights Group and wrote a report titled "Ad Tech Surveillance on the Public Sector Web" (Cookiebot 2019). In the report they explained that when users land on the website of the NHS in search of health data—and most families living in London in search of health information would land on the NHS website because it is the most trusted website by UK users—the NHS website would automatically send the data to multiple agents including companies like Google or Facebook.

These data-sharing agreements and business models can directly impact the life of families, and potentially children. In their report, the authors at Cookiebot (2019) highlight the implications of these data-sharing tactics by showing not only that health data can be used to infer a specific health condition or life situation, which can accompany an individual profile throughout a lifetime, but also that health data inferences can impact insurance scores. The report also shows that families have no clear way to prevent this information from being leaked, and they cannot correct or delete the data that has been shared or the assumptions that have been made about them.

Hence, we need to realize that under surveillance capitalism, there are clear privacy implications and real-life harm in health data web searches. We also need to realize that most of the data collected and processed through these searches in family life relates to the health of children. Companies are extraordinarily aware of this and design their business models often by targeting parents and children. One excellent example can be found if we look at Google. Google has been trying to harness health data searches for years now. This is because, according to their website, one in twenty user searches (worldwide, that is, millions of users) are health related. It is for this reason that the company has introduced features such as the Google Symptom Checker or the Google Knowledge Graph (figure 3.1). The first feature personalizes the search on the basis of symptoms; the second is a feature mostly to be associated with the app, so that when users search for health conditions they would get relevant medical facts right up front, which are approved by Google's team of medical experts.

What was fascinating about Google Knowledge Graph was that the feature is promoted by referring specifically to the image of the worried parent and children's health data. In a post on the official Google blog, the product manager writes: "... [M]y infant son Veer fell off a bed in a hotel in rural Vermont, and I was concerned that he might have a concussion. I wasn't able to search and quickly find the information I urgently needed (and I work at Google!)." Ramaswami then goes on to explain how he tried to find the solution to this problem by designing the Google Health Cards (Ramaswami 2015). Although his promotional blog post seems rather innocent, I believe that it is very interesting as it signals that Google is aware

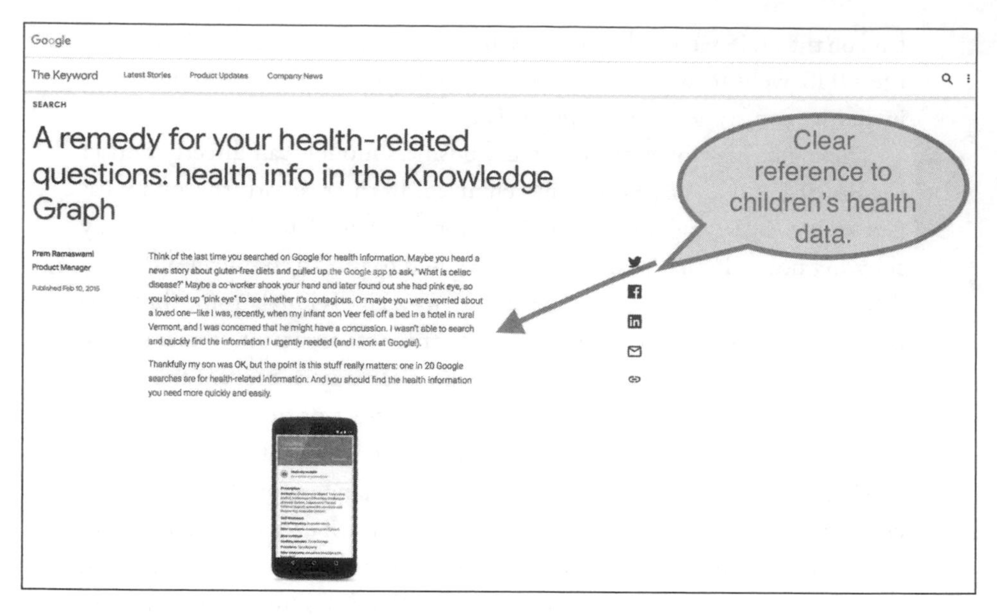

Figure 3.1
Google blog on the introduction of Knowledge Graph.

that much of the health data tracked through web searches related to the health of children.

Google is not the only web business who clearly knows that it is collecting the health data of children. Another example can be found if we consider a website like Babycenter.com or Babycentre.co.uk. Most of the families I met during my research, in both London and Los Angeles, would use the website mainly to search for health information about their children. Yet they were often unaware that, as mentioned in chapter 2, the website admits to "have a direct relationship" with ad technology partners such as Google, Amazon, AppNexus, and DoubleVerify. In the data policy, the company makes it clear that parents are responsible for their children's data flows. On their website they clarify that if "you submit any personal information relating to another individual to us, you represent that you have the authority to do so and to permit us to use the information in accordance with this Privacy Policy" (BabyCentre UK 2019).

As parents navigate these sites and rely on Dr. Google to find out details about their children's health, the health data of their children escapes their control and is collected, harnessed, and shared by a plurality of agents. The health data of children, however, is not only harnessed through web searches and websites, but also through other technologies such as mHealth apps, social media, and virtual assistants. Again, not only the Big Tech companies play a key role in the harnessing of this data, but data privacy implications for children are often dismissed by making the responsibility fall on parents.

mHEALTH TECHNOLOGIES AND CHILDREN'S PRIVACY

On an average day, families use a wide variety of health apps, wearables, and other technologies that broadly fall into the market definition of mobile health. The global mHealth market size was valued at $23.66 billion in 2017 and was estimated to expand at a compound annual growth rate of 25.7 percent between 2018 and 2025 (Grandview Research 2018). There is an extraordinary variety of mHealth technologies used in family life, including digital health apps (to communicate with health professionals); fitness tracking technologies (e.g., specific apps and wearables); nutrition apps, mental health apps, medicine apps; apps and wearables for chronic condition management (e.g., diabetes; heart, circulation and blood, among others); and of course conception, pregnancy, baby apps, and wearables as well as technologies designed for children.

One of the most groundbreaking and interesting research with reference to self-tracking apps can be found in the work of the sociologist Deborah Lupton (2016). In 2017 she also edited a special issue in the *Health Sociology Journal* in which she collected various contributions exploring the everyday use of mHealth and their social and implications in terms of power relationships, embodiment, and other aspects.

Of all the different mHealth technologies used by families, I am particularly interested in pregnancy and baby apps because these are the very first technologies that enable the datafication of children from before they are born (Barassi 2017a). In 2016, as mentioned in chapter 2, as I was awaiting ethical clearance to start my fieldwork, I conducted a small research project titled "BabyVeillance," which consisted of mapping the political economic environments of the ten most reviewed pregnancy/baby apps among UK and US users. The research was also based on an analysis of the promotional descriptions and data policies of the apps as well as on 3,570 reviewers' comments. Once I was ready to start interviewing and working with parents, I further developed the research through an analysis of parents' experiences and understandings of data tracking.

Pregnancy and baby mHealth technologies collect large amounts of health data of both the mother and child and share this data with a plurality of other companies, including Big Tech. Yet families are rarely aware of this. During fieldwork I met Maya and her wife Lindsay, who lived with their two dogs, one cat and their 6-month-old baby, in a small bungalow in a low-income neighborhood in Los Angeles. When I went to visit them, they offered me a Coke and we sat down on a rug in the living room while their baby crawled around and interacted with the dogs.

V: Do you use health-tracking apps?

M: I used the Ovia app to get pregnant and track ovulation. When I got pregnant I would use it to see what he was like? Is he a blueberry? Is he an apple? You know just to find out what to expect. I would count kicks, and things like that.

L: [Interrupting the interview] then we used the baby tracker, what is it called?

M: Yes, we used Sprout for tracking feedings and other things, and we were both connected to the same app, so if I took a nap she would know when he last fed, how much or when he pooped. But I did feel that it was actually more work to do than to just tell her. It was hard to put it down and record all that data. My brain just couldn't do it; the tracking I mean.

Despite Maya admitted that she stopped using her baby-tracking app to conceive her baby and in the nine months of her pregnancy, she dutifully tracked her health through the Ovia app. Since then she has never deleted it from her phone. When I studied the Ovia pregnancy app, I realized how Ovia's pregnancy app offers the ability to track multiple forms of health data of the woman and of the baby (figure 3.2). The data collected from the woman is divided in the following categories: ♦ *Weight— Symptoms— Nutrition— Medications & vitamins— Sleep— Moods— Exercise*. The data of the baby includes the following categories: ♦ *Customizable baby name, gender, goals, and health tracking, Pregnancy milestone tracking for belly pics, ultrasounds, baby shower, and more* (Ovuline Inc. 2019).

Considering the amount of health data that is produced and collected via an app like Ovia, we quickly realize that the data collected has the potential to be extraordinarily specific and granular. Companies are not only collecting the personal data of mothers and babies but they are turning it into profit

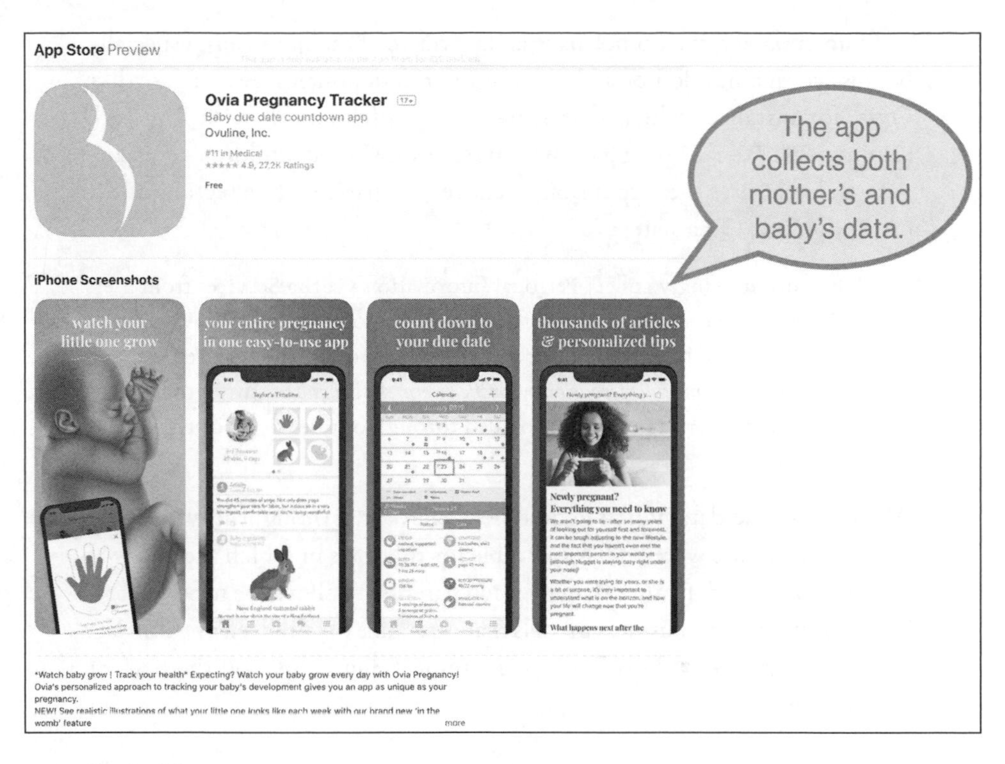

Figure 3.2
Ovia's promotional blurb in the app store.

by sharing this data with other companies. In 2017, the Electronic Frontiers Foundation published a report titled "The Pregnancy Panopticon" (Quintin 2017). The report demonstrated not only that pregnancy apps raised serious concerns for data privacy, but also that they shared personal data with Big Tech companies like Facebook, Google, and Amazon. In February 2019, the *Wall Street Journal* suggested that mobile apps shared information with Facebook Inc., including body weight, blood pressure, menstrual cycle, or pregnancy status (Schechner and Secada 2019).

As mentioned previously, that mHealth companies share data with Big Tech implies that this data can potentially be mapped to a unique ID profile. This could have critical implications for individual privacy and lead to real-life harm for the person involved. Hence key questions emerge in relation to children's data. Can children be identified? Will the information of children in the womb be available in their future?

Unfortunately, we do not have an answer to these questions, especially because companies do not address it in their data policies. A company like Ovia, for instance, which shares pregnancy information with Big Tech (Quintin 2017) does not engage with the question about children's data privacy and delegates the responsibility entirely to parents. In a section of the privacy policy, Ovia states:

> We do not knowingly collect Personal Information via the Services from users in this age group [beneath the age of 13]. We may, however, collect information about children and babies from their parents who are using our Services. *The parent or guardian assumes full responsibility for the interpretation and use* [my emphasis] of any information or suggestions provided through our Services for the minor. (Ovia 2018)

When I read the data policy I could not believe it. I thought it was deeply unfair that parents were held responsible for the ways in which the data was used and *interpreted*. Rather than blaming parents, I believe we need to challenge contemporary business models that capitalize on children's health data flows and start asking how companies are profiling people on the basis of their health.

HEALTH PROFILING: FROM SOCIAL MEDIA TO VIRTUAL ASSISTANTS

In summer 2017, Nicole was diagnosed with an autoimmune disease, and when she had her first infusion she posted a picture on Facebook. Her post was flooded with comments of support by family and friends. During the interview we talked about it and she explained why for her it was so important to post that health data on Facebook:

> N: I posted something about it [my first infusion], because I wanted people to know what I was going through, it is not only happy times that we post on Facebook, it's also times of sadness and struggle. And to put it out there in case other people are going through it, and they have experienced it so that I know I am not alone. When I posted the picture and explained what was going on, a lot of my friends understood me and it just made it feel like I wasn't so hopeless. When you think you are the only one, you think this is a terrible thing. When I first got my diagnosis and treatment, it seemed as all my life changed and I was mourning my whole life, and

> knowing that people had gone through that ... well it sucked that they had to go through that, but it made me realize that it was not only about me.

For Nicole posting about her health on Facebook mattered. It mattered because it made her feel that she was not alone. When it comes to health data, social media have started to play a fundamental role in bringing people together. This is evident if we consider an article published in 2014 in the *Health Informatics* journal in which a group of scholars reviewed sixty-three studies of social media–empowered patient practices (Househ et al. 2014). Social media can play a critical role in bringing people together who have to engage with specific health conditions or life changing events such as becoming parents.

During the research, I followed the Facebook parent groups that some of the participants of my research followed. The groups connected parents of specific local areas or brought together families with specific interests. For years, I made sure that the Facebook algorithm would read my interest for these groups and hence would keep me updated. Through online participant observation I noticed that a great majority of posts involved health data. Parents, and parents to be, asked advice about pregnancy, preterm labor, breastfeeding, and a variety of health-related topics from eczema, ear infections, and allergies to autism and developmental concerns.

As I witnessed these flows of health data, which were often associated to photos of children, I kept asking myself what would happen to that health data and whether that data could be used to profile children. Would Facebook be able to match the data with a face-recognition ID (which they call a template)? Would Facebook store that data and then add it to a user profile when a child—who was 13 years of age—opened a Facebook account? I personally could not find an answer to these questions, yet I was worried. In my worry I was not alone. Nicole for instance posted about her health but she did not want to post about her children's health, because she was concerned that health data on social media platforms could be used to profile them.

Although we do not have a clear answer as yet to what happens to children's health data flows (and also those of adults), there are clear indications that social media data is used by health and life insurers, and that more and more AI is being developed to bring together the data from different websites (especially social media) to profile individuals. In January 2019, for instance, the New York State Department for Financial Services granted life insurers the possibility to *evaluate* individuals on the basis of their social media posts to decide premiums, provided that the life insurance companies do not use that data to discriminate users (Scism 2019).

Considering how health data is being harnessed on social media, we also need to think about how social media data can be used to infer specific mental health conditions. Since 2006, for instance, Facebook has been working on suicide prevention and in the last few years it has launched a new AI for suicide prevention around the world (with the exception of the EU, because under the GDPR, companies are not allowed to profile users on the basis of sensitive information). The AI scans posts for patterns of suicidal thoughts and reports back to a team at Facebook, who then can decide to contact local first responders.

In an article that appeared on the *New York Times* on December 21, 2018, told the story of a police officer in Ohio who received a call from Facebook about a local woman who earlier that day had written that she intended to kill herself. The officer contacted the local woman who denied having suicidal thoughts, but she was told that she had to go to the hospital for a psychological assessment either voluntarily or in police custody, and he drove her to the hospital. The article does not tell us whether there was a real risk or not. Yet it highlights the complexity of the intervention program. It clearly shows that we need to be critical of a private company—that profiles users on the basis of their mental health and acts as a public office—without disclosing how it reaches its conclusions about users at risk (Singer 2018b).

I find the example of suicide prevention particularly insightful because it *does* seem to suggest that Facebook has the technology to scan social media posts on the basis of health-related content, and as Constine (2017) commented in *TechCrunch*:

> The idea of Facebook proactively scanning the content of people's posts could trigger some dystopian fears about how else the technology could be applied.... There are certainly massive beneficial aspects about the technology, but it's another space where we have little choice but to hope Facebook doesn't go too far. (Constine 2017, para. 5)

We cannot predict the future and find out what Facebook will actually do with the health data collected or how it will use it to profile individuals (and possibly also children). Yet what we know for sure is that under surveillance capitalism health data is used to profile individuals in often nontransparent and nonaccountable ways. Hence we really need to ask critical questions on the ways in which this data is used to profile individuals across a lifetime and what damage it could do.

These questions become particularly important if we consider how home technologies and virtual assistants might be starting to play a significant role in health profiling. In September 2018, the Google-owned Nest, which is known for its home technologies, bought the Senosis app, an mHealth app that focuses on diagnosis (Spanu 2018). Amazon was recently issued a patent (Jin and Wang 2018) for "Voice-Based Determination of Physical and Emotional Characteristics of Users." The patent will enable Amazon to develop a new technology to infer from voice patterns such as sniffles, crying, and other abnormal changes whether the user is sick or has an emotional issue, and then use that data for targeted ads.

This new patent aimed at the profiling of health becomes even more problematic if we consider that in April 2019, Amazon Alexa skills have become Health Insurance Portability and Accountability Act (HIPAA) compliant. The announcement was accompanied by the launch of six voice programs built by large health businesses ranging from Boston Children's Hospital, to the insurance giant Cigna, and to digital health company Livongo. The new tools allow patients to use Alexa to access personalized information such as progress updates after surgery, prescription delivery notifications, and the locations of nearby urgent care centers (Ross 2019). The question is whether Amazon will have access to this data and whether they will aggregate it with other forms of personal data into unique ID profiles.

CONCLUSION

Under surveillance capitalism, health data has become a big business; one of the most profitable and one of the most problematic. As this chapter has shown, Big Tech companies are harnessing large quantities of health data not only through their platforms and services but also by buying this health data from others. Furthermore, these companies are playing an active role in the datafication of the health sector by tapping into EHRs, introducing new AI and data infrastructures, and by cutting deals with health services. This implies that they have the potential to monitor the health life of individuals from the moment they are conceived to the moment they die.

What really concerns me about this transformation is the understanding that these companies have the means and technologies to match health data to unique ID profiles, and hence with other types of data, such as social media data, education data, or home life data. I believe that the mapping of health data with unique ID profiles constitutes one of the greatest privacy threats of surveillance capitalism. A lifelong profiling of individuals on the basis of their health could lead to all sorts of discriminations, inaccuracies, and biases.

The exchange of this data with governments and other agents can create the basis for new forms of inequality and breach of human rights, especially at the time of global pandemics. By the time that the COVID-19 pandemic started, I had already completed the book and the research. Yet I followed with interest and concern all the debates about contact-tracing apps and health tracking, and the growing role of tech companies in monitoring the pandemic. I was worried not only about the discriminatory potentials of COVID-19 health tracking but also about the data monopoly of Big Tech companies. As Deborah Peel, head of the US-based Patient Privacy Rights campaign, once mentioned, we also need to take into account the problem of data ownership (Khan and Lauerman 2018). These companies are turning patient data into secret intellectual property. This may make the treating of future patients very expensive and can exacerbate the already existing gap in health care between rich and poor.

4

EDUCATIONAL DATA: CHILDREN AS DATA SCORES AND SUBJECTS

When I was a child, I was terrified about grades. I studied and worked hard to make sure that I had a good grade average. I knew how much grades mattered; how much they defined me as an individual not only in front of my teachers and peers but also in front of my family. Fifteen years later when I had to choose my university, I carefully read the world rankings in my specific subject area. Once again I knew that data mattered; university rankings were going to play a fundamental role in my chances of getting a good job. When I became a teacher, I looked forward to student evaluations with anxious anticipation, and I have learned how to turn those evaluations into numbers; numbers that have come to define my professional performance. Grades, data, evaluations, rankings, and scores have always been part of my relationship with the education sector. That my embodied and lived experience of education has been directly shaped by data is not surprising. This is because, especially in Europe and the US, the history of the education sector cannot really be understood without considering the history of measurement (Lawn 2013).

It is by considering the history of measurement in education that we can appreciate why big data has had such a massive impact on the sector as a whole. The first part of the chapter will explore the impact of big data in education by discussing two different yet interconnected transformations that have occurred in both the UK and in the US over the last two decades. The rapid growth in usage of digital technologies in schools enabled the production of unprecedented amounts of educational data. At the same time, in the last decades in both countries the government and the private sector have facilitated the creation of educational data infrastructures and databases with large-scale financial investments (Williamson 2017a). Today these data infrastructures and databases are used to harness and process the data of individual students and institutions in ways that were not possible before, and the education system has become datafied.

As schools embrace new tech solutions and the demands of a data-driven education system, parents and children do not have any control over the data produced, and their consent is rarely taken into account. Yet under surveillance capitalism, the impact that educational data can have on a child's life is immense. In current data-driven educational environments, more and more public and private entities have access to personally identifiable student data. They collect it and use it to profile students, turn them into data scores, and make data-driven decisions about their lives. This situation is exacerbated by the largely deregulated and nontransparent market of educational data brokers, which sell educational data to potential recruiters, credit loans, insurers, and a plurality of other agents.

The data harm that can emerge from educational data scores or data sharing agreements can be profound. Like all the grades, evaluations, and rankings that I have encountered in my life, educational data can determine the life opportunities children have access to and can influence their sense of identity. It is for this reason that we need to critically consider the social and political implications that arise when this data is processed by privately owned algorithms. These nontransparent technologies rely on preemptive strategies of personalized learning, and in doing so may be locking children in stereotypes from a very early age and impacting their life opportunities.

EDUCATION: A HISTORY OF MEASUREMENT

The history of education cannot be understood without looking at the history of measurement. In the US in 1904, Thorndike, an educational psychologist at the Teachers College in Columbia (Lawn 2013), introduced the idea that education needed a "standardized unit of measurement." In 1924, Sears, an educational researcher in California, believed that a new philosophy of education was emerging that understood the importance of measuring. At the time, everything in education was starting to be measured. He listed cost, teaching efficiency, progress through school, success in studies, mentality, buildings, equipment, textbooks and attendance as some of the elements that were being measured (Sears 1924 quoted in Lawn 2013, 111).

The importance of measurement in education cannot really be understood without taking into account the historical context of early industrial capitalism and the broad understanding that different sectors of society (including education) needed to strive for efficiency. According to Lawn (2013), in the education sector the emphasis on efficiency was the direct result of a population and tax crisis that affected the US. Hence, measurements and control methods were introduced to try and reduce costs and to defend education against cutbacks.

Lawn (2013) believes that following the Second World War, the ideas of the measurement movement in education were then exported to Europe through international scholarly exchanges and through the creation of international education bodies such as the UNESCO Institute of Education in Hamburg. This is probably correct. In fact, if we want to understand the relationship between measurement and education, we cannot disregard international exchanges and the influence of the US measurement movement in education.

Yet we also need to be aware that measurement in education developed in Europe much earlier than in the 1950s. In England, for instance, educational psychologists such as Valentine, Burt, Isaacs, and many others had a fundamental influence in the education sector in the first part of the 1900s. They argued not only that children differed from each other in their innate intellectual abilities, but also that these intellectual abilities could be in fact measured (Wooldridge 2006). The understanding that human intelligence and abilities are innate and can be captured and measured is still of course very much shared among educators and psychologists in the UK today, who believe in psychological/psychometric assessments. Key examples can be found in the GL Assessments that are commonly used by schools in England such as the Pupil Attitude to Self and School (PASS) which is used to "identify fragile learners," or the CAT4 test for schools, which aims to assess students' "cognitive abilities" and personalize learning.[1]

The belief that we can measure an individual's cognitive ability or innate qualities has a long history and finds one of its most vivid representations in IQ tests. Yet this understanding is exceptionally problematic. As Gould (2006) argued in the book *The Mismeasure of Man*, we need to debunk simplified and positivist understandings that intelligence (as other aspects of human experience) can in fact be measured. We also need to realize that measurements are not objective, they are the product of cultural beliefs and values, and for this reason they are mostly biased.

It is within these (often biased) cultural beliefs in the power of measurement that we need to locate the cultural history of data in education. When we think about educational data and measurement, however, we need to realize—as Lawn (2013) suggests—that we are simultaneously talking about the measuring of teaching practices and institutions as well as the measuring of individual children and their differences. It is by simultaneously measuring, scoring, and evaluating the child and the institution that education can be governed at a national level and policy can be informed (Ozga 2009; Souto-Otero and Beneito-Montagut 2016).

BIG DATA AND EDUCATION

Understanding the history of measurement in education enables us to explain why the education sector is particularly prone to buy into the promises of big data. The rise of big data has brought about an overhaul of the sector in both the UK and the US. This radical transformation was made possible by a gradual change in educational practices, government incentives. and policies that happened over the last twenty years.

In the last decades in both countries, governments have pursued policies and strategies that led to the datafication of the education sector. They have imposed self-evaluation exercises on school administrators and teachers and have created data infrastructures that can measure and compare institutional performance. The aim was not only to determine whether an institution met national goals, but also to turn institutions into data scores that could be ranked against one another.

At the same time, the governments in both countries invested in the creation of longitudinal student databases that collect the personal information of students across their educational life. Between 1998 and 2002, for instance, the UK government launched the National Pupil Database (NPDB). According to the *Defend Digital Me* campaign, a UK not-for-profit organization that works to protect student data, in 2019 the NPDB counted 23 million unique student registries. This database contains the longitudinal educational data of students, who are also individually identified through their unique pupil number (UPN), which follows them across their educational life. Similar steps have been taken also in the US. Between 2005 and 2009 the federal government allocated $295 million to forty-one states who implemented Statewide Longitudinal Data Systems (SLDS), which allocated a unique ID number able to follow students as they progressed in the school system (US Department of Education 2009).

These sorts of initiatives in both the UK and the US signal to a historical transformation, which was fueled by the strong belief that governments needed to incentivize and support the creation of centralized data infrastructures for educational data. The shared assumption was that data infrastructures enabled those working in government, social work, or education to follow students across their educational life, measure their performance, and monitor their progress. The same data infrastructures also enabled policy makers to evaluate and monitor institutions and make sure that teaching standards were met in every school and district across the nation. Ben Williamson, author of *Big Data in Education* (Williamson 2017a) who works at the University of Edinburgh, once told me, "there's a huge layer of education politics that has developed over the last decades, which makes the datafication of children particularly concerning. This is because children became proxies for school performance, while schools became 'data centers.'"

THE BIG TECH IN SCHOOLS

The datafication of education came hand in hand with the digitization of the sector, as government officials, school administrators, and teachers were influenced by the belief in the potential of Ed Tech for the development of personalized learning strategies. Educators' fascination with digital technologies dates back to at least the 1970s, when small groups of educators across the US began to experiment with personal computers as tools for the development of a child-centered educational approach (Ito 2012).

In the 1980s and 1990s, these developments blossomed into a full educational movement fostered by educators and technologists (Ito 2012). At the heart of this movement lay the understanding that digital technologies *can* integrate learning and play, reinforce the participation of children and students who come from less privileged backgrounds, and allow a more personalized learning experience.

As director of the Connected Learning Lab at the University of California Irvine, Mizuko Ito shows that there is a lot of potential in Ed Tech education, and that digital technologies can transform the experience of learning. At the same time her work also demonstrates that Ed Tech education raises critical questions about how digital technologies can actually exacerbate existing social inequalities in education.

One of the main problems that we encountered in the last decades is precisely that—unlike Ito and her team—those working in Ed Tech often did not recognize the problem of digital equity and have been influenced by techno-optimistic and utopian dreams of educational technology (Selwyn 2013). This is particularly true if we consider debates concerning the introduction of big data technologies in schools. Mayer-Schönberger and Cuckier's (2014) *Learning with Big Data: The Future of Education*, is a fantastic example of the idealism and techno-optimism that defines current debates on big data in education.

After publishing their first and internationally renowned book on the big data revolution (Mayer-Schönberger and Cuckier 2013), the authors turned their attention precisely to education. They argued that big data is there to radically transform education. On the one hand, they believe that educators will be able to measure their teaching materials and practices and receive real time feedback, which will provide them with greater knowledge and insight into the teaching process. On the other hand, they believe that big data will make learning more accessible, by individualizing how knowledge is communicated to every single student and through other practices of personalized learning.

Mayer-Schönberger and Cuckier's (2014) belief that big data can create more accessible, engaging, and personalized forms of learning while enabling greater efficiency through data-driven decision making is shared by many in the education sector. It is for this reason that educational departments, institutional administrators, and teachers are opening up schools and institutions to different forms of educational technology. It is for the same reason that over the last ten years the educational technology market has expanded exponentially.

Although Apple and Microsoft have traditionally provided most of the technological services in schools worldwide for decades now, other companies like Facebook and Google are quickly finding their space in the market. Google in particular is gaining competitive advantage. The company started to take its first tentative steps in education around 2006 with the development of what later became the G Suite for Education. In 2019, thirteen years later, G Suite has 80 million users worldwide. In addition to that thirty million Chromebooks are now being used in education, and Google Classroom counts forty million users (Li 2019). In 2015 the Chan Zuckerberg Initiative, a company that was launched by Zuckerberg and his wife and that invests in education, health, and other areas, invested in Summit Learning, a personalized learning platform and technology, which at the time of writing is used by 74,000 students across the US (Bowles 2019).

As Big Tech and countless other educational technology companies (see Williamson 2017a) are rapidly allowed to enter schools, we need to engage with critical questions about the implications of these transformations. Selwyn (2013), for instance, in *Distrusting Educational Technology*, openly criticized the techno-optimism that defines the relationship between technologies and education. The book shows that much of the promise of tech education is rooted in a neoliberal educational model that favors student–computer interaction over other forms of interaction, and business needs over creative and unpredictable learning practices. The book also shows that the introduction of privately owned education technologies in schools can erode the public nature of education and reinforce individually centered forms of learning. Most of his points are, at the time of writing, being raised by students and parents who are concerned about the introduction in schools of platforms such as Google Classroom or Summit Learning.

One of the fundamental questions that emerges when we think about how schools and children's learning experiences are being transformed by educational technology is the question about data. In *Big Data in Education*, Williamson argued that schools are being turned into data centers always assessing their own performance through data points, while students are becoming the subjects of increased data mining and analytics (Williamson 2017a).

Children today are becoming the subjects of educational data mining practices because their data is being used for educational data science, which is based on the belief that by collecting as much data as possible about the person and the context in which learning unfolds, educators can predict future outcomes and improve the learning environment as a whole. This implies not only that student data is collected but that students are profiled on the basis of this data so that educators can intervene. These forms of interventions are preemptive in nature and risk locking children in stereotypes that can affect their lives and impact their future opportunities. What exacerbates the problem is that the profiling of students is undertaken by for-profit technologies, whose algorithms are protected by trade secrecy (Williamson 2017b) and whose accuracy is open to question.

PERSONALIZED LEARNING? BETWEEN PREDICTION AND PREEMPTION

On November 14, 2018, one hundred Brooklyn students walked out in protest against the implementation in their schools of Summit Learning, the online education program that promotes personalized learning and that has been funded by the Chan Zuckerberg Initiative. The students complained that their learning experience was being dominated by screens; they also believed that they were being left alone in the learning process. This was not the first protest of students against Summit Learning or the last. In 2017, following protests by parents and students in Cheshire, Connecticut, the program was terminated. In April 2019, different schools in Kansas revolted against the use of Summit (Bowles 2019).

When students in Brooklyn took to the streets to protest against Summit Learning, Hernandez and Robison, two students who organized the protest, wrote a letter to Mark Zuckerberg explaining their reasons:

> Unfortunately we didn't have a good experience using the program, which requires hours of classroom time sitting in front of computers ... the entire program eliminates much of the human interaction, teacher support, and discussion and debate with our peers that we need in order to improve our critical thinking. Unlike the claims made in your promotional materials, we students find that we are learning very little to nothing. It's severely damaged our education, and that's why we walked out in protest. ...
>
> Another issue that raises flags to us is all our personal information the Summit program collects without our knowledge or consent. We were never informed about this by Summit or anyone at our school, but recently learned that Summit is collecting our names, student ID numbers, email addresses, our attendance, disability, suspension and expulsion records, our race, gender, ethnicity and socio-economic status, our date of birth, teacher observations of our behavior, our grade promotion or retention status, our test scores and grades, our college admissions, our homework, and our extracurricular activities. Summit also says on its website that they plan to track us after graduation through college and beyond. Summit collects too much of our personal information, and discloses this to 19 other corporations. What gives you this right, and why weren't we asked about this before you and Summit invaded our privacy in this way? (Quoted in Strauss 2018)

Hernandez and Robison's letter was published in the *Washington Post* (Strauss 2018), together with the response by Raymonde Charles, the communications director for education at the Chan Zuckerberg Initiative. In his response, Charles made it clear that Summit Learning did not have any intention to sell student data and that the company adheres to the Student Privacy Pledge, a binding legal commitment introduced in 2014 by the Future of Privacy Forum and signed by the major educational technology companies.

Yet if one reads Summit Learning's privacy policy what is concerning is not really whether the company shares or sells the educational data, but how the company stores and uses this data to profile students. On its privacy policy, Summit Learning makes it clear that it is gathering the different types of student data mentioned by Hernandez and Robison's letter. This is the data that they openly admit collecting:

> Student name; Email address; Student identification numbers such as school identification number or school information system; Student record information such as attendance, suspension, and expulsions; Student demographic data, including date of birth, gender, ethnicity or race, and socioeconomic status; Student subgroups, such as English learners; Student subgroups, such as students with disabilities; Mentor observations; Student outcome information such as grade level promotion and matriculation, AP and IB test information, college admission test scores, college eligibility and acceptance, and employment; and Academic or extracurricular activities a student may belong to or participate in. (Summit Learning 2019)

The company clearly articulates that they use all this data to "provide curricula choice or recommendations" or "drive learning engagement and progress via content suggestions to users." Hence we need to ask ourselves, What does it mean to live in a world where students can be profiled on the basis of their social status or race? And is this data then used to target them with specific educational content within their schools?

Although I understand why educators can be attracted by the promise of personalized learning offered by companies like Summit Learning, we need to realize that personalized learning cannot really happen without profiling children. In addition we need to understand that the shared assumption of companies like Summit Learning is not only that students' abilities can be scored and students can be profiled, but also that through data we can have a clear picture of their learning necessities, and hence design their educational programs accordingly.

Profiling is sold to educators as a way to take action and mitigate future risks in learning. Thus we cannot really understand profiling in education without considering the notion of *preemption*. One of the promises of predictive analytics and profiling in the age of surveillance capitalism is precisely the idea that through data we can preempt problems. This understanding is particularly influential in predictive policing (Dencik et al. 2017) or anti-terror operations (Elmer and Opel 2008). Individuals are profiled as potential criminals through data-driven decisions so that police can take action before an event happens.

The problem with interventions that are of a preemptive nature is that as my colleague and friend Lina Dencik, codirector of the Data Justice Lab at Cardiff University, told me, these interventions suggest social stasis and that people are locked into stereotypes. This is particularly concerning when we think about children and education. By profiling children and offering them personalized content, programs like Summit Learning may be stereotyping children, pigeonholing them in specific positions and stalling their right to social mobility. The mission of education should be precisely the opposite—ensuring that children have access to many different areas of knowledge, and that they can experiment with all these areas in open-ended and non-discriminatory ways.

THE PROFILING AND SCORING OF CHILDREN

Once we understand that personalized learning can lock students in specific stereotypes, which can affect their educational opportunities and life chances, we can turn our attention to the fact that the profiling of children is carried out by for-profit organizations which use nontransparent algorithmic models. Consequently, we cannot be sure that the profiling of students and the actions that the school takes to prevent possible problems are accurate and in their best interest.

One of the most interesting examples in this regard, I believe is represented by the baseline assessment, which will be introduced by the UK government in primary schools in England from 2020. The assessment is to be carried out in the first weeks of reception—which is the equivalent of the US kindergarten—on children between 4 and 5 years old. One of the most contested and contradictory aspects of the assessment is that it is meant to produce a unique number, which scores and ranks the child and represents a baseline figure against which the progress of the child is going to be measured throughout primary education years. What is particularly shocking about the baseline assessment is that it does not have any educational use to teachers—its main purpose is to hold schools accountable for children's progress.

In a detailed report on the implications of the baseline assessment, Bradbury and Roberts-Holmes (2016) studied the 2015 pilot launch in schools and identified fundamental problems with it. They have shown that the baseline assessment had a negative impact on the first—and very important—introductory days of school. That teachers were forced to turn students into data scores deeply affected the process of induction and bonding that is vital to a child's first days of school. In addition to this, their report shows that the teachers and educators struggled with the idea that children could actually be defined and profiled by a single unique number and did not find the assessment particularly insightful.

What I found particularly interesting in the report was the finding that teachers and educators questioned the accuracy of the private providers that ran the assessment. At the time of the pilot, the baseline assessment was run by three different providers, and all the different providers offered different formats of assessments and different results. This left the principals and teachers questioning the accuracy of their models, and "feeling vulnerable to a largely deregulated and nontransparent private sector" (Bradbury and Roberts-Holmes 2016, 45).

The example of the baseline assessment in the UK shows that key problems and critical questions emerge when schools try to reduce children to a data point. It also shows that in building a just and fair education system, governments cannot and should not be informed by data scores that are actually provided by private companies, which rely on secret and unaccountable algorithms.

The baseline assessment in the UK is one of the many examples of the ways in which children are being profiled through privately owned and nontransparent algorithms. Another example can be the application of facial recognition in schools. As Andrejevic and Selwyn (2020) have shown facial recognition technologies are being used for different reasons in schools: for security reasons and possibly to identify school shooters (this is especially true in the US), to monitor attendance, to check students' identity using online platforms, or to determine student (non)engagement on the basis of a facial analysis. One of the main problems with these applications is that, as Harwell (2018) has shown in an article that appeared in the *Washington Post*, not only facial recognition technologies remain unproven technologies (especially as deterrent to school shootings), but the surveillance firms that provide schools with these technologies remain unaccountable for the ways in which they use children's data and profile children. Another problem with these applications is that facial recognition technologies are used to create specific psychological profiles of students, which reduce learning to a passive process of knowledge consumption (Saltman 2016) or an individualized psychological behavior change exercise (Williamson 2017b).

As parents, civil society organizations, and educators start reflecting on the implications of student data privacy and profiling, one of the key questions that emerges is the issue about data harm and how the profiling of children on the basis of their education data could come to affect their lives in significant ways. The harm that educational data can cause in children's lives is clear if we look at educational data sharing agreements and data-brokering practices.

EDUCATIONAL DATA HARM

In both the UK and the US between 2011 and 2015, we have seen a relaxation in laws regulating educational data in order to facilitate data disclosures to third parties. In the US in 2012, an amendment to the then existing Family Educational Rights and Privacy Act (FERPA)[2] allowed for greater disclosure of student identifying information (Mendelsohn 2012). In 2013 in the UK, an amendment to the Education Individual Pupil Prescribed Persons Regulations 2009 was introduced to allow access to third parties to individual pupil information (Defend Digital Me 2019).

The disclosing of educational data to third parties is raising critical questions at the grassroots level about who these third parties are, how they share information, and what harm these sharing agreements can actually cause. In the UK since 2012, for instance, children's sensitive personal data has been released from the NPDB more than 1,000 times (Defend Digital Me 2019). Most of the time, the pupil level data that was collected within the national database has been distributed for free following explicit requests. Yet these requests and disclosures have become a cause for concern among civil society organizations and privacy groups especially because it became clear that student personal information was shared with government departments that caused harm and distress to children and their families (Defend Digital Me 2019).

During the Child | Data | Citizen project, I had a clear example of how the data harm created by data exchange agreements could be possible. In January 2017, I went to pick up my daughter from her preschool in the heart of a big social housing estate in south London. Although the area had been in fact gentrified over the last ten years, it is still largely a low-income neighborhood and the home to a variety of ethnic minorities and social housing estates. My daughter had been at that day care/preschool since January 2015. We had originally chosen that particular location not only for convenience—in fact the day care was close to our home and on my way to work—but also because when we first visited it, we immediately fell in love with its caring atmosphere.

That day, I came back from work eager to see P and rang the bell as always. Charity, the manager, told me about the events of the day and asked about my work. Then, out of the blue she told me:

C: "Remember tomorrow to bring in P's passport, I need a copy of it."

I paused; I looked at her and asked "why?" and she added

C: "Oh it's just for our records, it's the new policy of the company [that runs the day care]."

I didn't have much time to ask questions. It was late, my daughter needed to go home for dinner, and I was tired and pregnant. Yet that evening I started to think about the implications of that request. It was the beginning of 2017, just seven months after the Brexit vote. I was living as a European citizen in the UK at a historical time that was not only defined by the uncertainty of Brexit, but also by the establishment of the "hostile environment" a set of legislative and administrative measures announced by the Home Office in 2012 under the lead of Theresa May to tackle illegal immigrants in their schools, their homes, and workplaces (Usborne 2018).

As a European citizen who had been living in London for 16 years, I was very well aware that my rights were no longer to be taken for granted after Brexit, and I was particularly sensitive to the numerous stories that emerged in mainstream media about the Windrush scandal. The Windrush scandal affected those migrants who had the right to settle in the UK as they came from British colonies during the 1960s and 1970s, but were not given any documents to prove their rights. At that time, especially the migrants that came from the Caribbean (the Windrush generation) were being wrongly detained and deported by a system that left little scope for discussion or rectification of errors. Parents were being separated from their children, and families were being destroyed in front of my eyes after they had been living in the UK for 50 years. Every time I read a new story about the irrationality of the *hostile environment* I felt angry and terrified.

When Charity asked me to bring P's passport in, I decided to dig deeper. I found out that the practice of recording pupils' birth data and nationality had been introduced by a guidance of the UK Department of Education in 2016, which stated that schools needed to seek birthplace data from parents, although it was not compulsory for parents to provide it. I also found out that by December 2016 the Department of Education had a data exchange agreement with the Home Office to share information of up to 1,500 pupils a month (Gayle 2016) and that schools were known to be requesting the data mostly from non-white non-British children (Whittaker and Camden 2016). That evening as a I did my research, I felt angry and worried. I was not worried about P, because she is a British citizen, but I was very well aware that about 80 percent of the children at the preschool were non-white, and many were non-British.

The next day I confronted Charity about the issue and told her that she did not have the obligation to ask that information. She explained to me that she had no choice, "that her hands were tied," that the decision came from "above" and that it was a general policy, which was implemented at national level by the company that owned the preschools. I knew that it was a battle that I should have pursued further, but I was 4 months pregnant, alone in London with P who was 3 years old, and I barely managed to cope with the multiple responsibilities of my working and family life. So the next day I brought P's passport in and Charity dutifully scanned it.

Thankfully, other charities, unions, and activists fought the struggle for me, and after sustained criticism, in 2018, the UK Department for Education announced that schools in England were no longer encouraged to ask evidence of pupil's nationality and pupil's country of birth. This example however shows that when we think about student data privacy, a lot is at stake, and this is especially because we have very little knowledge of the multiple ways in which a student's data is shared and used. The 2016 data exchange agreement between the Home Office and the Department for Education in England was of course rectified, but the very existence of these data exchange agreements in the first place demonstrates the profound social and political implications that are intrinsic to student data flows and what happens when student data is accessed by third parties.

If we want to raise critical questions on student data harm we cannot only focus on the harms that emerge with data exchange agreements, such as the 2016 agreement between the Home Office in the UK and the Department of Education. We also need to consider the harm produced by the largely deregulated market of educational data brokers. At the time of writing, student data privacy laws are mostly aimed at schools and online vendors of educational software, but do not target data brokers. The world of educational data brokering is still largely deregulated, and parents, students, and educators, have no awareness or control of what is going on beneath the surface.

DATA BROKERS AND THE MARKET OF STUDENT DATA

In 2018, the Center for Law and Information Policy at Fordham Law School published a shocking report titled "Transparency and the Marketplace for Student Data" (Russell et al. 2018), showing that educational data brokers in the US operate largely deregulated, and that they sell student lists with personally identifying details that profile students on the basis of ethnicity, affluence, religion, lifestyle, awkwardness, and many other categories. The report identified fourteen data brokers that admitted to selling student data.

According to the report, companies such as Accurate Leads, which sells lists of information on kids in high school—and claims to have 9,895,179 student names—publicizes that buyers can purchase access to the following personal data: "Age with parents [sic] name, student by class year with their name, gender of child, parents age, household income, propensity to buy specific products and services, net worth, lifestyle factors, own vs. rents, length of residence, marital status, ethnic [sic], parents [sic] education level and much more." (Russell et al. 2018, 10). Hence the ways in which data brokers are actually selling this data is through the systematic profiling of children on the basis of family, class, ethnicity, gender and many other personal and sensitive categories. The researchers at Fordham for instance asked a data broker if they could send them a list of "fourteen and fifteen-year-old girls" for family planning services and the data broker agreed to send them the list (Russell et al. 2018, 3).

Data brokers are selling the data of children as young as 2 years old (Russell et al. 2018), but unfortunately it is very difficult to have a clear picture of what data brokers are actually doing with young children's data or the way they collect it or share it. However, we can start grasping some of their practices if we look at how they harness and tap into high school and higher education data. In 2018, for instance, the *New York Times* broke the news that the data collected through college-planning surveys completed online by around 3 million high school students in the US was being sold by data brokers.

One example reported in the article is the example of Scholarships.com, an online platform that allows high school students to search for scholarships. The platform asks students for their name, birth date, race, religion, home address, and citizenship status and whether they have "impairments" like HIV, depression, or a "relative w/Alzheimer." All this data is then shared with Scholarship.com subsidiary company known as American Student Marketing (ASM) (Singer 2018a).

The company is identified in Russell et al. (2018) as one of the operating educational data brokers in the US. When I visited their website in May 2019, they claimed to offer a listing of "high school, college bound high school, college, college graduates, grad school, students, parents, young professionals, workforce and households data." They also suggest that marketers can search "students and parents by home and email address, academic and artistic interests, athletic interest and involvement, student organizations, honor societies and awards, ethnicities, gender, GPA/SAT/ACT, demographics, geographic information and much more."

According to their promotional website, marketers can use the list "for outreach via postal direct, email mobile, text, digital social media campaigns and more." I was intrigued and decided to click the section on high school students. On the ASM page (figure 4.1), the company identifies the type of businesses or marketing campaigns that the list can be used for, and in their list they include student credit cards, student loans, and trade and career education institutions.

The data harm that can arise from these trades is self-evident. Students (and their families) are being profiled through the harnessing of sensitive personal data, gathered from a variety of sources, such as the online survey that they completed to search for possible scholarships. These profiles are then sold not only to commercial brands but also to trade and career education institutions, student credit cards, and student loans. This could impact their life choices and opportunities.

Student data-brokering practices are common also in the UK. In 2014, the UK Universities and Colleges Admission Service (UCAS) made 12 million

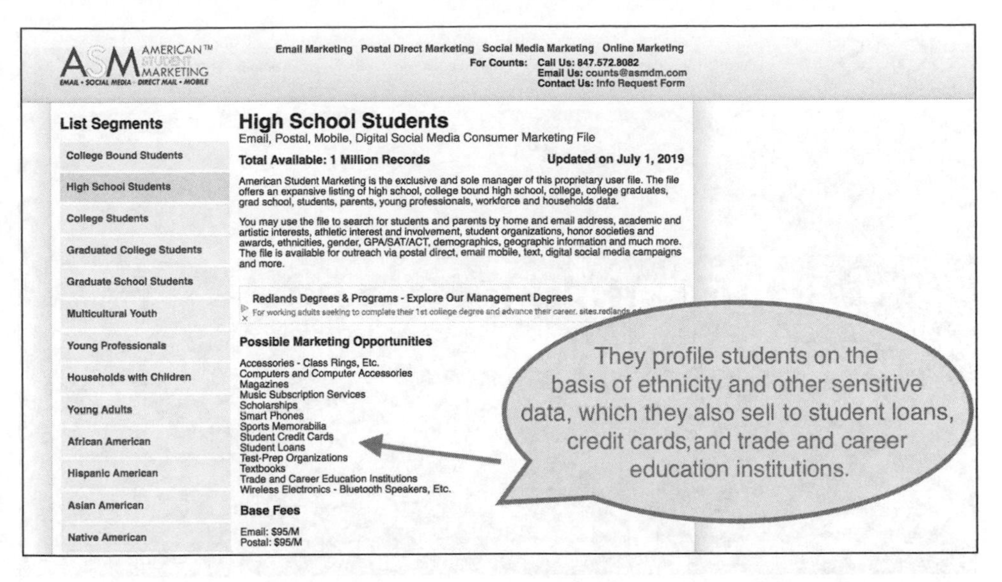

Figure 4.1
American Student Marketing (http://www.asmdm.com/).

in revenue by selling student data (Ward 2014). When I visited their UCAS Media—Student Marketing and Recruitment Solutions website in May 2019, which is aimed at advertisers and recruiters, the organization openly publicized that they can help recruiters to "identify the right talent" and "strategic insights."

As figure 4.2 shows, UCAS advertises its products to different parties: education providers, employers, media agencies, commercial brands, and news. In the section on commercial brands, they promote themselves in the following way:

> We know which students already have a place confirmed at university, where they're studying, and what they're interested in. We understand their decision-making behavior, and we'll work with you so your message aligns with their journey. This will have more impact, give them what they need, and give you better returns. (UCAS 2019)

Although such forms of marketing practices may seem a great business solution, we need to start addressing the social and political implications of businesses practices that profile children and youth on the basis of their personal data and in doing so may be denying them the access to key life opportunities.

Figure 4.2
Universities Colleges Admission Services Media website.

As schools become more and more datafied and digitized and students turned into data scores, as educational data is collected and shared in ways that are still largely unknown to parents and students, it is important to look at the lived experience of families and to ask critical questions about consent and participation and what parents can do to protect themselves and their children.

THE PROBLEM OF PARENTAL CONSENT AND THE IMPORTANCE OF COLLECTIVE ACTION

In 2005 in the UK, Pippa King, the mother of two children aged 6 and 7 years discovered that her children were nearly fingerprinted without her consent for a school library system. She decided to find out why she was not informed or why the school did not ask for her parental consent. She confronted the principal of the school, who told her that there was no legal obligation from the school to ask for parental consent. The then UK Data Protection Act 1998 did not consider biometric data as sensitive personal data, so schools in the UK could store and process a child's biometric information without parental consent.

Since 2005, Pippa King has been campaigning, along with other parents and privacy/civil liberties organizations, for the regulation of biometric data in the UK education system. Her campaign "Biometrics in Schools" aims to ensure that children's biometric data is used in schools only with parental consent and that school provide alternative provision to those parents who do not want their biometrics used. This is not an easy task; from 2001 to the present, biometric technology in UK schools has included the use of children's fingerprints, iris scans, facial recognition, and infrared palm scanning for a range of activities such as library books use, canteen payments, locker access, and registration (King 2018).

Pippa King's personal experience as a parent, as well as her campaign work on biometrics in schools, demonstrates how fragile, problematic, and complex is the issue of parental consent when it comes to education data. The State of Data report (Defend Digital Me 2019), of which Pippa King is one of the contributors, clearly articulates the problem of consent in schools. The report shows that when it comes to educational data flows, children are entirely at the mercy of the decision making of school staff, who are often overworked and undertrained to understand the data privacy implications. It also shows that parents are rarely asked for their consent.

I had a clear example of this during the Child | Data | Citizen project. Parents I worked with, including myself, were experiencing a rise in use of outsourced platforms and apps in schools, and schools introduced these technologies without offering a space for debate or asking parental consent. The introduction and use of data technologies in schools was often presented to parents as an inevitable and inescapable reality.

During the project, I talked to different parents who felt that they had no choice or control over education technologies in schools. Angela, for instance, who was a single mother of an 8-year-old enrolled in a north London school, told me that she was sent an email by the principal of her school, informing parents that the school would introduce Google Classroom to be used for homework. She also told me that different parents replied to the email raising different concerns, which highlighted three potential problems with the platform:

1. The alleged opening of an email account,
2. Increased screen time for children, and
3. Children's data privacy.

All concerns were dismissed in a response email from the principal. I had confidential access to this email and was struck by the fact that it left little space for debate and offered no alternative. Here is a rough summary of the response from the principal:

1. Email account: The email stated that children would not be given an email account so there was no reason to worry.
2. Screen time: The email stated that the school expected that the homework would take between ten and twenty-five minutes, depending on the age of the child and the activity. It also said that "how much time parents allowed their child to spend on screens was not a problem for the school to deal with."
3. Data privacy: The email stated that Google Classroom abides by the General Data Protection Regulations (GDPR) and thus there was no reason for concern.

The response given by the principal was dismissive and did not really address parents' concerns. Although children do not have an email account they usually have a unique identifiable Google account. The opening of a unique ID profile should be debated with parents, and parents were right to raise their concern.

The principal dismissed any privacy concern by mentioning that "Google complied with the GDPR and so there was nothing to worry about." Yet, being in compliance with data regulations does not mean that the technologies that children use are immune to privacy risks; it just implies that parents have been informed and have given their consent to the collection and processing of their children's data. This is precisely why the parents had written to the principal raising their privacy concerns—they wanted to make an informed decision. Unfortunately, their concern was not addressed by the response of the principal.

Yet what really bothered me when I read the email was that it offered no alternative to parents. In the email, the principal explained that one of the reasons for choosing Google Classroom "was to streamline teacher's work and be more efficient with time and resources," and hence the school "did not offer an alternative homework." The email also stated that "the school was no longer going to provide paper copies of homework," and that if parents wanted to print those copies they could, but that "all copies of homework had to be submitted on Google Classroom."

Angela's experience of the introduction of Google Classroom was not isolated. A similar experience was shared also by Cara, who was a single mother of a 10-year-old and lived in Los Angeles. Although Angela did not really mind the introduction of Google Classroom and did not give much attention to the email of the principal (although she decided to share it with me), Cara was annoyed about the lack of choice and debate in her school when it came to using Google services for education. In her interview, she told me

C: I am a bit annoyed about the implementation [of Google Classroom]. My daughter created her account at school and told me when she came home. If they put it in a paper I didn't see it. Plus I don't get why she got access to a Google drive; they did not have my permission.

V: Did you say anything?

C: No. I reached out one time in the past because they shared a YouTube link that was public, but they found an excuse and I worry that I care too much and I don't want to come across as annoying. I told them not to post about her, she is in less pictures but in some of them she is. They also give her printed instructions to sign up for quizzes or things like that online. They are not really tech-savvy, and recently they told us that we should be on YouTube, they are giving us instructions to use websites that are not privacy sensitive. We didn't have this debate at school.

As the experiences of Angela and Cara show, schools are implementing technologies and platforms; they are allowing Big Tech corporations to gather and store massive amounts of educational data without opening up the debate to parents or addressing the problem of educational data harm. The COVID-19 crisis has exacerbated these practices and accelerated this situation, and I experienced this firsthand, as I found myself agreeing to the terms and conditions of Google Classroom and opening a profile for P. As more and more parents find that they cannot opt out, there are questions that emerge when we think about how these technologies use children's data to profile them and target them with learning content. These questions need to be addressed by parents, schools, and businesses, who need to join forces and act together to protect student data privacy.

CONCLUSION

The history of the education sector in the UK and US cannot really be understood without considering the history of measurement and data. Yet the rise of big data has radically transformed the role of data in education. Today, technologies, algorithms, and visualizations are becoming such a fundamental way in which we have come to define schools and children, and we have come to rank them against one another. Data is seen as holding the key to the understanding of children's innate abilities and schools' structural problems. Data is also seen as the solution to addressing these problems.

At the heart of this sociotechnical transformation in education lies the belief that data in education can allow for the development of personalized forms of learning and enable data-driven decision making. It is for this reason that in the last decades we have seen a steep rise in the use of educational technologies in schools. It is for the same reason that we have also witnessed large-scale financial and political investments in the creation of educational data infrastructures that can track schools and monitor individual students across a lifetime.

As the education sector becomes more and more datafied, we need to ask critical questions not only about student data privacy, but also about the harms that emerge through the harnessing and profiling of educational data. Educational data flows, which are largely deregulated and based on often unknown data sharing agreements, not only raise fundamental questions about student data privacy but also about algorithmic explainability and accountability. We need greater transparency to guarantee that educational data is not used to discriminate students in ways that lock them into specific stereotypes and impact their life opportunities.

Parent's collective action in schools is now needed more than ever. Parents like Cara cannot feel alone and powerless when it comes to challenging the introduction of data technologies. Something needs to be done, and we should praise and welcome political initiatives such as the European Union proposed ban of facial recognition technologies in public spaces for five years.

5

HOME LIFE DATA: HOME TECHNOLOGIES AND CHILDREN'S RIGHTS

In autumn 2016, I was carrying out one of my first interviews for the Child | Data | Citizen project with Julie, the mother of a 4-year-old son who lived in Los Angeles and had just bought Amazon Echo. She was amused by her son's interaction with the voice-operated AI. She told me that when her son came back home from preschool he would sit down next to the speaker and taunt Alexa:

J: He tells her very silly things like "you are such a goof" or "you are stupid" or "you are dumb" and things like that, and she replies: "Well you are not so nice." So, he comes to me saying that Alexa told him that he was not very nice. He is really excited about it. I think for a small kid, it's exciting. There is this voice that is coming out of this box and responds to his commands.

V: But does he understand that Alexa is not real?

J: He thinks it's real. Like with Google Maps. He is convinced that there is a person there telling us where to go. Also, with Alexa, he thinks that there is a person that is talking to us who responds. You know Alexa is not dumb; it's an AI program. So, he is fascinated by the fact that he can have a conversation with her.

As voice-operated AI become part of everyday living, children like Julie's son talk to them, ask questions, make jokes, test their intelligence, or make specific requests. When we think about the domestication of artificial intelligence and home automation, there are so many fascinating questions that emerge about machines and cultural values (Crawford and Calo 2016), about our understanding of intelligence (Broussard 2018) and shifting notions of what it means to be human. In the context of family life, it would be important to explore how human behavior and family relationships are being transformed by these technologies or what it means for children to grow up with artificial intelligence devices that provide them with the illusion of humanity and teach them to respect and relate to an AI/object as a quasi-person (Elgan 2018). Unfortunately, during the Child | Data | Citizen project I did not have the time or resources to engage with these questions. What I had the time to do was to look at the question about data privacy.

The issue of data privacy in home automation is quickly coming to the fore. In April 2019, the news broke that a team of Amazon employees were listening to Alexa's recording (Day et al. 2019). A month later, in May 2019, it transpired that Amazon had been granted a patent that enabled Alexa to process the conversation that occurred before the wake word (Piersol and Beddingfield 2019). Hence it became clear that Alexa recorded conversations even when users were not aware of it. In July 2019, the Belgian broadcaster VRT NWS got access to 1,000 leaked recordings of Google Assistant, which included "bedroom conversations, conversations between parents and their children" as well as "professional phone calls containing lots of private information" (Morse 2019, para. 5–6). The broadcaster also revealed that most recordings happened without the Hey Google command and hence without users knowing that they were being recorded.

All of these instances highlight the deep-seated privacy implications that are involved when we think about the inclusion of virtual assistants into family homes and the issues that emerge when we think about children's data privacy. As a society we are just starting to negotiate with these technologies and to understand their intrinsic threats, especially for children. At the time of writing we are witnessing the emergence of the very first lawsuit that addresses the problem. In June 2019, Amazon was sued for allegedly recording children without consent (Kelion 2019).

This chapter will explore the ideologies and business models that shape home technologies, as well as their privacy implications for children. By introducing the concept of *home life data* (Barassi 2018) and discussing the issue of household profiling, this chapter will show that the data produced within the home is not only personal/individual data, but it is also data about the family as a social group, its socioeconomic context, values, and everyday messy behaviors. The privacy implications of a technology that can integrate highly contextual family data with biometric data such as a voice or face prints are immense, and we need to strategically think about the data harm that can emerge for children's lives.

HOME AUTOMATION: FROM A HISTORICAL IDEA TO A PRESENT REALITY

The idea of the automated home, a type of home that is structured and organized by intelligent technologies, has a long history and dates back to the discourses of modernist architecture. Le Corbusier's famous claim that the house is a "machine for living in" (Timmerman 2007) stressed the analogy between technology and architecture: homes needed to be functional and essentialist.

Between the 1920s and 1940s, different corporations and institutes, such as General Electric and Massachusetts Institute of Technology (MIT), sponsored projects that tried to connect technologies to domestic life (Chambers 2016). In the 1950s, the first example of a technological home, where almost any task could be completed by pushing a button, was created. The home, known as the Push Button Manor, was built in 1950 by the inventor Emil Matthias in Jackson, Michigan. These early steps in the field of automated homes were followed by the inclusion of futuristic home visions such as the "home of tomorrow" that developed in the 1960s or the techno-utopian discourses of science fiction literature (Aldrich 2006). Toward the end of the century, internet-connected technologies have significantly amplified the possibilities for home automation, giving rise to a new understanding of the meaning of *connected* and *technological home* (Chambers 2016; Strangers 2016; Aldrich 2006).

Even if we can trace the early developments of home automation back to the last century, it would be misleading not to appreciate that the development of domestic artificial intelligence devices is taking the concept of home automation to a new dimension. In the last couple of years, the home automation market has grown exponentially and is predicted to continue to grow. A report published in January 2017 by Juniper Research has estimated that smart home hardware and service, which include entertainment, automation, health care, and connected devices is set to drive revenues from $83 billion in 2017 to $195 billion by 2021 (Juniper Research 2018).

At present we are seeing the emergence of a wide variety of home automation technologies that are transforming our homes, which include artificial intelligence devices (e.g., virtual assistants, robots that act as home assistants; artificial intelligence toys); energy and utilities (i.e., companies that utilize sensors, data to monitor water and energy consumption); security technologies (e.g., smart locks that allow users to replace home keys with locks and talk to guests; surveillance cameras that are operated through apps; alarms, which can detect intruders and are equipped with special sensors to detect floods, fires, and so forth); home appliances (e.g., smart refrigerators, smart toilets); entertainment devices (e.g., smart TVs, whole house wireless music systems; video games); lighting monitoring devices (e.g., smart bulbs and switches that can be controlled at a distance); and unique solution devices (e.g., devices that offer different specific solutions, such as support with recycling or intercom solutions).

All the different technologies mentioned above are not an exhaustive list of the multiple ways in which home automation technologies are transforming the home. The pace of technological transformation and the extensiveness and pervasiveness of the developments in the Internet of Things (Greengard 2015) makes it incredibly difficult to compile such a list. Each day we see a new start-up, a new business trying to enter the market. Yet what the list above suggests is that presently there is an incredible variety of technologies that are becoming significant players in family homes.

The list also suggests that one of the challenges that businesses are facing at the moment relates to the fact that in order to build a truly automated home, all the different technologies need to communicate with one another. In December 2016, in a Facebook post called "Building Jarvis," Zuckerberg described precisely this problem. He discussed how he tried to build an AI for his home, which he called Jarvis after the butler in the film *Iron Man*. The post showed that, at the time, it was very difficult to build an AI that created a truly automated home because there was a shortage of common standards and Application Programme Interfaces (API) that allowed different brands of smart technologies to talk to each other.

In the last few years, however, something changed. We witnessed the development of a new business model for home automation and domestication of artificial intelligence brought forward by the Big Four companies: Amazon, Google, Apple, Samsung (and other companies, like Microsoft and Facebook are trying to establish themselves as key players too). According to the Juniper report, the Big Four at present dominate the smart home market and will further solidify their position, with Amazon securing a leading role (Juniper Research 2018). These companies have created a business model for home automation, which I will refer to as "home hubs," that extends across different techno-social dimensions. One way in which we can explore these dimensions is by referring to the idea of "information ecology."

THE BIG FOUR AND THEIR BUSINESS MODEL

The concept of information ecology was introduced by Nardi and O'Day (1999) to describe that when we study technologies we need to consider how these technologies are shaped not only by technical structures but also by the interaction between people, practices, values, and discourses. This implies that we need to take into account how technology is shaped by specific human and economic relations as well as human beliefs and practices. The concept also implies that we need to be aware that "different parts of the ecology *coevolve*, changing together according to the relationships in the system" (Nardi and O'Day 1999, 1).

During the Child | Data | Citizen project I analyzed four home hubs—Amazon, Google, Apple, Samsung—by looking at their business models (e.g., economic relations and platform structures), their promotional cultures (the discourses that they promote, how they see themselves), and their data policies. The analysis revealed that the different home hubs rely on complex information ecologies, and that the changing relation between the different elements of each ecology is dictated—unsurprisingly—by market competition.

The business model of home hubs is quite complex and is structured (broadly speaking and at the time of writing) by four different dimensions. The first dimension is of course the *AI virtual voice assistant* (Amazon Alexa, Google Assistant, Apple Siri, and Samsung Bixby). Virtual assistants are usually operated by home speakers. (However, as mentioned by Professor Leah Lievrouw from UCLA, during a chat on data and privacy in 2018, these are not only "speakers" but also recording technologies.) Virtual assistants can be integrated into a variety of home technologies and beyond (especially if companies have an open platform model). The AI assistants operate through speech recognition and are connected to specific profiles and accounts (e.g., Amazon and Google) but also enable multiple users to add more family members to a unique profile.

The second dimension of the business model is defined by the different *services* that users can access through the assistant. In very simplistic terms, we can understand these services as "voice-operated apps" that families can access through the interaction with their virtual assistant (e.g., Alexa Skills, Google Actions, Siri Shortcuts, Bixby Commands). These services are continuously expanding. For instance, to regain competitive advantage over its competitors,

Apple developed Siri Shortcuts by tapping into its 2 million apps (Casinelli 2018). In order to extend Alexa Skills, Amazon created the Alexa Fund, which provides up to $100 million in venture capital for companies that build Alexa Skills Kit. Between 2016 and 2018, Alexa Skills have increased from 5,191 in November 2016 to 30,006 in March 2018 (Kinsella 2018).

The third dimension of the business model of home hubs is the development of *smart home technologies*. Samsung has always been at the forefront in the production of smart home technologies (e.g., such as the Samsung smart fridge, which is basically a family hub). Apple has also invested in the design of specific smart home accessories for the HomeKit. In 2014, Google/Alphabet acquired Nest, which produces different home technologies (such as thermostats and security technologies). The Alexa Fund invests in home technologies that are produced by others yet are compatible with Alexa and in 2018 it also signed a deal with the Lennar Corporation, which is building 35,000 automated homes in Florida, which are all operated by Alexa (Weise 2018).

The fourth dimension of their business model is defined by mobile home apps that enable control of the home remotely from a smartphone (Alexa App; Google Home app; Apple iOS Home app; Samsung Smart Home app). It is through the app that companies can have a full picture of family life, collect large quantities of behavioral data, and follow their users.

THE MYTHS OF HOME AUTOMATION

The four different home hubs do not only rely on similar business models but also on similar promotional cultures that emphasize three different, yet interconnected technological myths (Mosco 2004): *techno-solutionism*, *data fetishism*,[1] and *AI quasi-humanity*. In his groundbreaking book *The Digital Sublime*, Vincent Mosco argues that the Western fascination with the "newness" of technologies has enabled us to construct technological myths, which make us perceive technologies as if they were almost "magical" and socially transformative (Mosco 2004). One of the greatest myths that we find in Western technological thought is the belief that technologies are "the solution to the most complex problems" in our lives and our societies (Morozov 2011, 2013).

The myth that technologies are the solution to all life problems is strongly emphasized in the promotional cultures of home hubs. The Big Four promote their home hubs precisely by relying on this cultural understanding.

Amazon and Google encourage users to "Just Ask Alexa" or "Just say Hey Google" and promote the idea that as a solution to different daily problems "all you need to do is ask." Apple and Samsung promote the idea that with Siri and Bixby you "get things done." Despite a difference in jargon, what is striking about the promotional cultures of home hubs is that they all emphasize the idea that virtual assistants are *the solution* to everyday life problems and can save you time, precious time, that you can spend with your family.

In the promotion of their home hubs, techno-solutionism meets another technological myth of our times: data fetishism. According to Beer (2019), in the age of surveillance capitalism, data is seen as the solution to make us more efficient and to have greater control of our lives, business, and society. This "imaginary," he believes plays directly to our fears—our fears of not being productive or efficient enough, of being left behind, or of not being in control of our lives.

In promoting their home hubs the Big Four stress this data imaginary by referring to two ideas: control and mobility. For Amazon "you can control your home, anytime, anywhere with Alexa," for Apple you have "your home at your command." Samsung refers to "the ultimate control of your home," and Google mentions often the terms "command" or "control," but doesn't seem to have a short promotional blurb to convey this understanding. What is fascinating about the emphasis on control and command is that this emphasis is often attached to a form of data fetishism. In fact all the four hubs promote the idea that it is important to monitor, track, and keep note of everyday mundane aspects of family life in order to have control of our lives.

As Humphreys (2018) has shown, documenting (and sharing) mundane details of our lives has historically been a human practice long before the advent of information and communication technologies. Yet the obsession with data production—that very idea that we need to archive and store everything—is a discourse that is particularly strong today and that says a lot about the current appeal to self-track (Lupton 2016). The business models of home hubs seem to emphasize and encourage self-tracking and self-monitoring by advancing the idea that the more data you have the better you can control your life.

There are multiple examples of data fetishism on the companies' websites and the website of their partners. One example that I loved when I was doing the research was the promotional blurb of Baby Stat, a Google Assistant "action," which promoted itself by suggesting that "Babies like stats!" Of course, babies don't like statistics, because they don't know what statistics are. Yet I think that the promotional blurb is interesting as it sheds light on the fact that in our data-obsessed environments, data is constantly fetishized. We cannot analyze the myths of home hub technologies without referring to the myth of artificial intelligence and the fact that voice-operated AI is promoted as quasi-human.

THE HUMANITY OF VIRTUAL ASSISTANTS?

The myth that we can build "thinking machines" that are able to emulate the cognitive faculties of humans dates back to the 1950s (Natale and Ballatore 2017). At the heart of this myth lies the understanding that if we build machines that think like humans, then we might be able to build machines that are almost human (discussed further in chapter 9).

The promotional cultures of all the four home hubs stress this myth by presenting virtual assistants and home technologies not only as thinking machines, which are also talking machines, but also as machines that *can think about you* and *care about you*. Amazon, for instance, encourages users to "enjoy a home that takes care of you." Home hubs users can usually set up personalized routines (e.g., a set of actions, such as lighting, music, or heating that can be activated with a unique voice command) so that their homes can take care of them when they come back from work or sit down for a family dinner. The overall idea is that your home and your virtual assistant can take care of all your needs from the moment that you wake up to the moment that you go to bed. The images on the websites seem to emphasize precisely this level of intimacy and care.

A common discourse that is emerging among the Big Tech at the moment and that reinforces ideas of the *humanity* in AI is the very understanding that AI virtual assistants are offering something extraordinarily new in terms of human computer interaction: an emotional interaction. According to Zuckerberg (2016), for instance, that we can talk to AI virtual assistants implies that we are seeking more emotional depth with our technologies. In his post, "Building Jarvis" he notes: "Once you can speak to a system, you attribute more emotional depth to it than a computer you might interact with using text or a graphic interface" (Zuckerberg 2016, para. 26).

In emphasizing the emotional connectedness that people feel with voice-operated machines, Zuckerberg was not alone. A survey carried out by

Google/Peerless Insights on 1,642 users of voice-activated speakers showed not only that people are engaging with these technologies as if they were quasi-human by saying "please," "thank you," and even "sorry," but also that 41 percent of people said that voice-activated virtual assistants feel like talking to a friend or another person (Kleinberg 2018). Of course, there are many anthropological questions that emerge about the context of such responses, questions that cannot really be solved by surveys. Numbers alone rarely speak about context, intention, and the human passions, desires, and contradictions that define our interaction with virtual assistants.

Although we do not know whether business developers are right that virtual assistants are really transforming our emotional interaction with technologies, what is becoming obvious is that this understanding is largely influencing the promotional cultures of home hubs, which make these technologies particularly appealing for families. Who wouldn't want a home that understands all our needs and solves problems for us or a virtual assistant with whom we could emotionally connect?

Yet as families buy into the promise of the automated home, they are required to give up massive amounts of highly contextual data. These data flows can impact not only children's right to privacy but also their right to self-expression and nondiscrimination within the home (Barassi and Scanlon 2019).

HOME HUBS AND PRIVACY IN FAMILY LIFE

In autumn 2016, when I interviewed Julie, she was very aware that Alexa was gathering and processing her personal information. When I asked her to imagine how much data and what data is collected from her family, she replied, "I have Alexa, so I think every minute. Every time that my son is talking to Alexa there is data being collected. . . ." Although Julie was aware that her Amazon Echo recorded what they said, she told me that she was not too concerned about her family's privacy because Alexa did not have the means to integrate data from other sources such as bills or social media accounts. Then she added, "You know it's different if there are different places that have different insights about you and not only one platform that knows everything." Of course, three years later, Alexa has become a very different technology from the one Julie described in 2016, and I wondered how she felt now about its privacy implications. Did she stop using Alexa?

What impact did the privacy scandals have on her? I wrote to her but I am still waiting for her to respond to my email.

During my research, I had the chance to talk to other parents who used home technologies. These parents, in contrast to Julie, were concerned for their privacy and the privacy of their children. Cara, for instance, who lived in Los Angeles, was a single mother who I introduced in the previous chapter. She told me that she did not "talk to Alexa much," precisely because she was concerned about privacy. Also Luke, who lived in London and had two children aged 1 and 4 years, told me that he would use his Amazon Echo only for "functional tasks" such as music and lighting, and that in the family the interaction with Alexa was very limited.

During the project, I also interviewed parents that resisted the trend to buy a home hub, precisely because they were concerned about everyday surveillance. Scott, the father of two daughters aged 8 and 10 years and lived in Los Angeles, refused to buy an Amazon Echo even if his wife and daughters wanted one because "It records everything and a lot can go wrong." Also Claire, who lived in London with her children aged 3 and 1 years, refused to buy a home hub because she was concerned about what the company might do with all that home data it collected:

> C: Now the trend seems to suggest that everything is automated, and you have everything on your phone, even if you are on holiday you can switch off your heating. You are mobile, you can do everything online. But all the data is there, stored somewhere. Your pictures, everything. So in a way, the positive side is that this gives you flexibility, and mobility, but the downside is what they actually can do with that information. I think that we need more regulations to limit what companies can do with home data, so that we can protect our rights and the rights of our children. I think we are just at the beginning; we are just starting to think about what can be done with all the information that is out there.

More parents are becoming aware that home hubs rely on a business model that is aimed at monitoring, collecting, and processing large amounts of personal data. To tackle their concerns, some parents like Cara choose to limit the information they provide to these technologies and to use them tactically (although these tactics might not be enough, because home technologies often record conversations even without the wake word). Other parents—like Scott or Claire—choose not to buy them. As more families reflect on the privacy implications of home hubs, we need to critically break down the type of data that these technologies collect and explore the implications especially with reference to children's rights.

HOME LIFE DATA AND CHILDREN'S PRIVACY

The question about home hubs and children's privacy has been tackled over the last few years especially by focusing on technologies that were designed for children. In 2017, Mattel cancelled its Aristotle AI assistant for kids amid privacy concerns. In 2018, we witnessed a growing debate from lawmakers in the US about Amazon's use of children's data in the Amazon Echo Dot for Kids. In 2019, the *Campaign for a Commercially Free Childhood* has launched a new investigation into Echo Dot for Kids that has shown that Amazon retained children's data also after parents tried to delete it.

Although it is clear that home hubs designed for children constitute a threat for their privacy, it is important to understand that one of the greatest problems that we face today in the automated home is that children often interact with home hubs and other home technologies that are not designed or targeted at them (Barassi 2018). That these technologies are not designed for or targeted at children implies that they do not have to abide to the Child Online Privacy Protection Act (COPPA) or to the GDPR's special regulations for children. Hence protecting children's rights in the automated home has become particularly problematic, especially because home technologies that are not designed for children do not explain how they gather and harness children's data.

Home hubs are directly collecting children's data through aggregated profiles, such as Amazon Household or Google Family Link, which enable different profile IDs of family members to come together. During the Child | Data | Citizen project, I conducted a small auto-ethnographic exercise on Amazon's Household to try to understand what happened to the data of children. Amazon Household is an interesting case study because it highlights the complexity of home data and how different individual profiles can be grouped together.

When I first studied Amazon Household in 2018, the company allowed the aggregation of a maximum of six individual profiles under a unique "household": two adults and up to four children. In 2019, the feature was extended, enabling the aggregation of up to ten individual profiles, which included the profiles of adults, teenagers, and children. My personal, auto-ethnographic journey as a parent of two to find more about the Amazon Household's privacy policy has been confusing and frustrating. After carrying out an ethnographic exercise in summer 2018 on Amazon UK (see Barassi 2018), I have carried out the same exercise in summer 2019 on Amazon US. The two exercises revealed that it was impossible to fully understand what happened to children's data. What follows are the five steps that I took to understand what happened to children's data; by Step 5 I gave up:

STEP 1
I land on the Amazon Household page (figure 5.1). It tells me that I could add under a unique profile two adults and up to four children and four teens. Yet there is no mention of compliance to COPPA or GDPR or any note to what happens to children's data.

Figure 5.1
"Manage your Amazon Household" page.

STEP 2

I start looking for a Privacy Policy on the page and, with great difficulty, I find it at the very bottom (figure 5.2).

STEP 3

I want to find out how to connect my Household profile to Alexa (figure 5.3). The page doesn't mention COPPA but tells me that children can access content on FreeTime. I can't have FreeTime on all my Echo devices otherwise my home automation won't work. This implies that my children will be interacting with Alexa beyond FreeTime. Yet there is no information about what Amazon does with these interactions.

STEP 4

I decide that the best chance I have to find out what happens to children's data is actually to search for the keyword "children's privacy." I end up on Amazon's Children's Privacy Disclosure, and although in contrast to the past (Barassi 2018) the company recognizes that it collects data from children's profiles, it is impossible to have a clear understanding of what happens to that data simply by reading their policy, especially because of the issue of third parties. In fact, Amazon clearly states that the disclosure does not apply to third parties (figure 5.4).

STEP 5

I give up. I feel deeply frustrated again and find myself agreeing with the fieldnotes I wrote in 2018:

> I don't understand, I feel so incompetent and frustrated. I have been reading the privacy policy again and again but fail to understand it. It

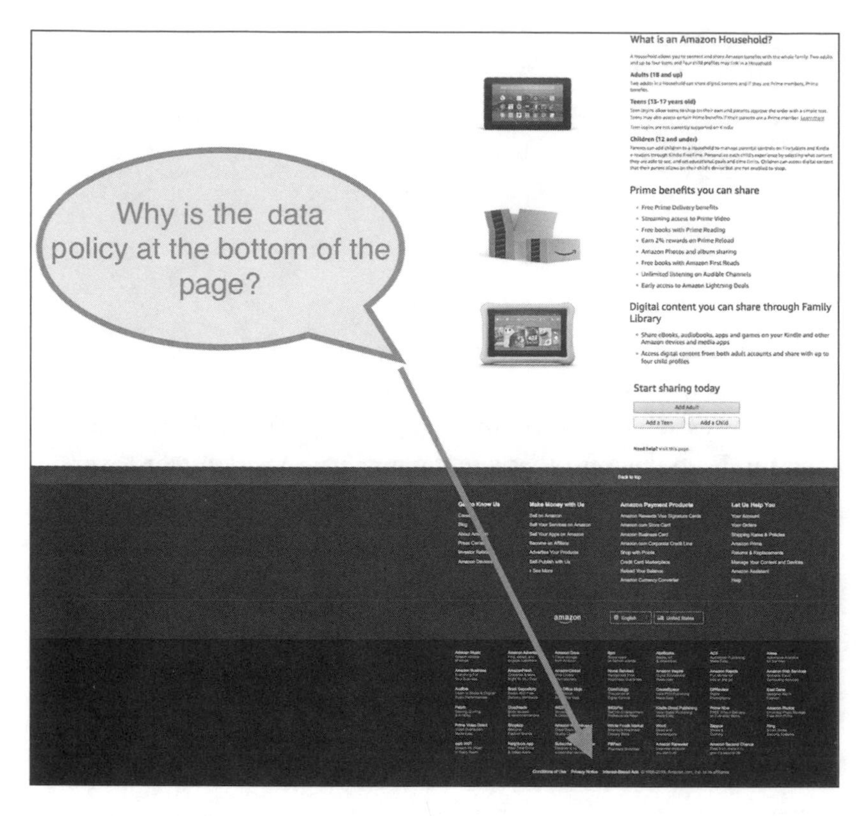

Figure 5.2
"Manage Your Amazon Household."

> is clear that the company recognizes that children interact with the virtual assistants or can create their own profiles connected to the adults. Yet I can't find an exhaustive description or explanation of the ways in which their data is used. ... I can't tell at all how this company archives and sells my home life data, and the data of my children. (Barassi 2018)

My short auto-ethnographic exercises made me reflect on the fact that we do not know how children's data is collected or harnessed from home hubs that are not designed for them. Through the aggregation of adult and children's profiles, home hub technologies have the potential to harness precious

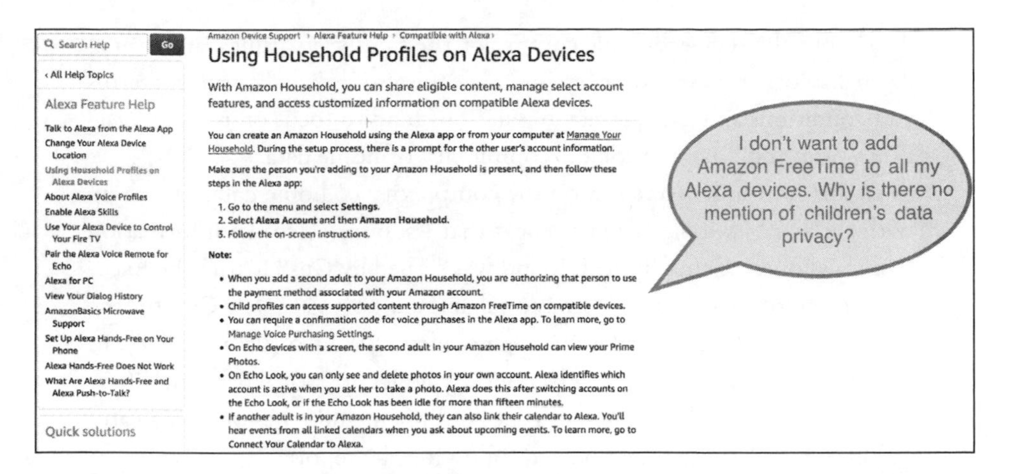

Search Help Go

‹ All Help Topics

Alexa Feature Help

Talk to Alexa from the Alexa App
Change Your Alexa Device Location
Using Household Profiles on Alexa Devices
About Alexa Voice Profiles
Enable Alexa Skills
Use Your Alexa Device to Control Your Fire TV
Pair the Alexa Voice Remote for Echo
Alexa for PC
View Your Dialog History
AmazonBasics Microwave Support
Set Up Alexa Hands-Free on Your Phone
Alexa Hands-Free Does Not Work
What Are Alexa Hands-Free and Alexa Push-to-Talk?

Quick solutions

Amazon Device Support › Alexa Feature Help › Compatible with Alexa ›

Using Household Profiles on Alexa Devices

With Amazon Household, you can share eligible content, manage select account features, and access customized information on compatible Alexa devices.

You can create an Amazon Household using the Alexa app or from your computer at Manage Your Household. During the setup process, there is a prompt for the other user's account information.

Make sure the person you're adding to your Amazon Household is present, and then follow these steps in the Alexa app:

1. Go to the menu and select **Settings.**
2. Select **Alexa Account** and then **Amazon Household.**
3. Follow the on-screen instructions.

Note:

- When you add a second adult to your Amazon Household, you are authorizing that person to use the payment method associated with your Amazon account.
- Child profiles can access supported content through Amazon FreeTime on compatible devices.
- You can require a confirmation code for voice purchases in the Alexa app. To learn more, go to Manage Voice Purchasing Settings.
- On Echo devices with a screen, the second adult in your Amazon Household can view your Prime Photos.
- On Echo Look, you can only see and delete photos in your own account. Alexa identifies which account is active when you ask her to take a photo. Alexa does this after switching accounts on the Echo Look, or if the Echo Look has been idle for more than fifteen minutes.
- If another adult is in your Amazon Household, they can also link their calendar to Alexa. You'll hear events from all linked calendars when you ask about upcoming events. To learn more, go to Connect Your Calendar to Alexa.

Figure 5.3
Using Household profiles on Alexa.

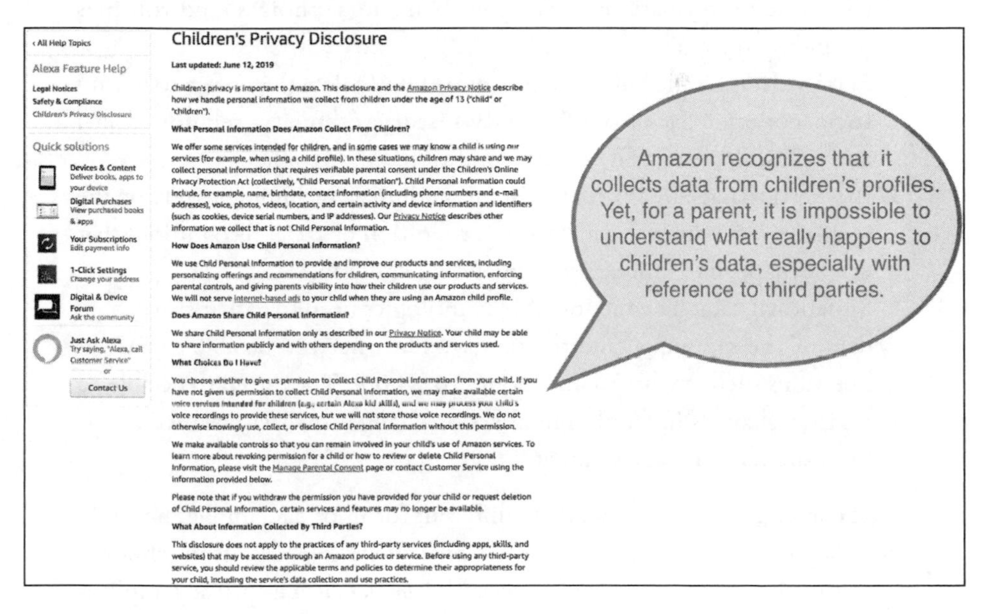

‹ All Help Topics

Alexa Feature Help

Legal Notices
Safety & Compliance
Children's Privacy Disclosure

Quick solutions

Devices & Content
Deliver books, apps to your device
Digital Purchases
View purchased books & apps
Your Subscriptions
Edit payment info
1-Click Settings
Change your address
Digital & Device Forum
Ask the community
Just Ask Alexa
Try saying, "Alexa, call Customer Service"
or
Contact Us

Children's Privacy Disclosure

Last updated: June 12, 2019

Children's privacy is important to Amazon. This disclosure and the Amazon Privacy Notice describe how we handle personal information we collect from children under the age of 13 ("child" or "children").

What Personal Information Does Amazon Collect From Children?

We offer some services intended for children, and in some cases we may know a child is using our services (for example, when using a child profile). In these situations, children may share and we may collect personal information that requires verifiable parental consent under the Children's Online Privacy Protection Act (collectively, "Child Personal Information"). Child Personal Information could include, for example, name, birthdate, contact information (including phone numbers and e-mail addresses), voice, photos, videos, location, and certain activity and device information and identifiers (such as cookies, device serial numbers, and IP addresses). Our Privacy Notice describes other information we collect that is not Child Personal Information.

How Does Amazon Use Child Personal Information?

We use Child Personal Information to provide and improve our products and services, including personalizing offerings and recommendations for children, communicating information, enforcing parental controls, and giving parents visibility into how their children use our products and services. We will not serve internet-based ads to your child when they are using an Amazon child profile.

Does Amazon Share Child Personal Information?

We share Child Personal Information only as described in our Privacy Notice. Your child may be able to share information publicly and with others depending on the products and services used.

What Choices Do I Have?

You choose whether to give us permission to collect Child Personal Information from your child. If you have not given us permission to collect Child Personal Information, we may make available certain voice services intended for children (e.g., certain Alexa kid skills), and we may process your child's voice recordings to provide these services, but we will not store those voice recordings. We do not otherwise knowingly use, collect, or disclose Child Personal Information without this permission.

We make available controls so that you can remain involved in your child's use of Amazon services. To learn more about revoking permission for a child or how to review or delete Child Personal Information, please visit the Manage Parental Consent page or contact Customer Service using the information provided below.

Please note that if you withdraw the permission you have provided for your child or request deletion of Child Personal Information, certain services and features may no longer be available.

What About Information Collected By Third Parties?

This disclosure does not apply to the practices of any third-party services (including apps, skills, and websites) that may be accessed through an Amazon product or service. Before using any third-party service, you should review the applicable terms and policies to determine their appropriateness for your child, including the service's data collection and use practices.

Figure 5.4
Amazon's Children's Privacy Disclosure.

data about the socioeconomic context of the family, its values, and everyday behaviors. Yet my short exercises made me also realize that current terms and conditions and data privacy regulations tend to focus on *personal (individual) data* and do not tackle the complexity of home data.

In 2018, in order to reflect on the complexity of home data, I came up with the term *home life data* in a report that I submitted to the Information Commissioner's Office in the UK and was signed by Gus Hosein, the Executive director of Privacy International and supported by Jeff Chester the director of the Center for Digital Democracy in the US (Barassi 2018). In the report I argue that home hubs not only collect personal data but different types of data. These different types of data include the following categories, which I revised and updated following the report:

1. Household data: Home hubs and smart technologies collect a wide variety of household data from shopping lists to energy consumption and gather key information on families' behaviors, choices, and routines (including the ones of children).
2. Family data: Home hubs gather a lot of family data that refers to family socioeconomic background, family history, ethnicity, religion, social and political values, and medical conditions among others.
3. Biometric data: Most virtual assistants and smart technologies rely on the gathering of biometric data (voice recognition or facial recognition) that can be mapped to unique users, including children.
4. Situational data: To function, AI technologies need to gather situational data of the individual and the family. They need to be able to answer questions such as what room belongs to whom. They need to be able to register changes in family members or changes in circumstances, conflicts and tensions, and so forth.

That companies can gather all these different forms of data implies not only that they have the potential to harness *highly contextual* data from children but also to integrate this data with *biometric information*. The privacy implications of technologies that can integrate context and biometrics are immense.

CHILDREN'S HOME LIFE DATA: BETWEEN BIOMETRICS AND CONTEXT

One of the critical questions that I kept asking myself when I was trying to navigate the data policy of Amazon Household was whether home hub companies were able to map all this home life data to voice prints and therefore build unique ID profiles of children. A voice print, like a face scan or fingerprint, is a uniquely identifiable piece of data. Once that data is linked to an individual it can be used to infer many things about them, and these inferences can follow children across a lifetime (Barassi and Scanlon 2019).

So far there is evidence that companies are developing technologies to infer sensitive information about users on the basis of their voice data. As mentioned, Amazon was recently issued a patent (Jin and Wang 2018) for "Voice-Based Determination of Physical and Emotional Characteristics of Users," which will enable the company to infer from voice patterns such as sniffles, crying, and other abnormal changes whether the user is sick or has an emotional issue, and then use that data for targeted ads. The question at heart for children's privacy is whether a company like Amazon can effectively recognize children through their voice prints, infer sensitive data about them, and use this data to profile them. Unfortunately, we do not have an answer to this because privacy policies and regulations do not address the problem of aggregated profiles or biometric data.

Yet what is becoming increasingly clearer is that through the aggregation of adult and children's profiles and the collection of household, family, and situational data, home hubs can gather and harness highly contextual data of children. The following examples are particularly illustrative (although a bit dated) of the ways in which developers are thinking about context:

> In 2015, Wulfeck one of the developers of the AI toy Hello Barbie mentioned in an interview on the *New York Times* that: "[Hello Barbie] should always know that you have two moms and that your grandma died, so don't bring that up, and that your favorite color is blue, and that you want to be a veterinarian when you grow up." (Wulfeck, ToyTalk in Vlahos 2015)

> In 2016 also, Zuckerberg wrote about the importance of context in his post on home automation. "Understanding context is important for any AI. For example, when I tell it to turn the AC up in "my office," that means something completely different from when Priscilla tells it the exact same thing. That one caused some issues! Or, for example, when you ask it to make the lights dimmer or to play a song without specifying a room, it needs to know where you are or it might end up blasting music in Max's room when we really need her to take a nap." (Zuckerberg 2016)

Context is key for voice-operated AI to work properly, and it is for this reason that companies need to gather highly contextual data. Yet a technology that is able to integrate highly contextual data with biometric data can have dramatic consequences on children's lives. It could be used to make wrong assumptions about them, to discriminate them, and cut them off from important life opportunities. This becomes evident if we consider how tech companies and data brokers are tapping into the promise of "household profiling."

HOUSEHOLD PROFILING: IMPLICATIONS AND THREATS

Household profiling is an established practice among data brokers and tech companies. It is for this reason that education data brokers, as shown in the previous chapter, sell not only the data of individual students, but also the data about their parents (job, ethnicity, financial situation, lifestyle factors, marital status, and so forth) (Russell et al. 2018). It is for the same reason that tech companies are trying to gather as much data as possible about the household. In 2018, Facebook, for instance, filed a patent request for the development of a technology titled "Predicting Household Demographics Based on Image Data" (Bullock et al. 2018). The technology would enable Facebook to "build more information about the user and his/her household" and to provide improved and targeted content delivery to the user and the user's household (Bullock et al. 2018).

As companies develop new technologies for household profiling we need to realize that these technologies are extraordinarily open to inaccuracy, fallacy, and discrimination. Families often do not use technologies as they are designed to be used. This is not only because on an average family day technologies and profiles always overlap and this confuses algorithms, but also because families often input inaccurate data in their technologies to use them tactically. This human messiness of home life data inevitably confuses algorithms and leads to inaccurate profiling.

For instance, during the research I had a personal auto-ethnographic example of this messiness. As mentioned elsewhere, Google knew that I was pregnant of A before my family knew and started targeting me as a future mom. Yet eight months into my pregnancy, my sister found out that she was pregnant, so I started looking for information online about her pregnancy. In the coming months, I realized—from the targeted ads that I was receiving—that there was a great confusion in the ways I was being profiled. The data sniffers on the web could not tell if I just had a baby or whether I was still pregnant.

Understanding the messiness of home life data is of central importance as it enables us to ask critical questions about the ways in which families are being profiled on the basis of these inaccurate and messy traces, and the harm this inaccuracy could bring about. In my report on Home Life Data and Children's Privacy (Barassi 2018), for instance, I make a simple example about inaccurate health profiling:

> Let's imagine, that I am having dinner with my friend who has a child who suffers from diabetes and I might ask my Google Assistant or Alexa to look for information on "diabetes in children." Let's also imagine that in the weeks to come I am concerned about my child getting diabetes and I start looking for information on symptoms. All these data traces would imply that I probably would be profiled as "parent" (which I am already profiled as) with a "diabetes interest." When we think about household aggregated accounts such as Amazon Household or Google Family Link, however, the question, emerges naturally: if I am inaccurately profiled as a parent with a diabetes interest, and my child is connected to my same household account would he/she be profiled as possibly diabetic? How is this data shared and collected by data-brokers? Will this inaccurate data impact on my child's life? (Barassi 2018, 5)

When we try to understand the potential impact of household profiling we cannot only focus on the problem of inaccuracy, we also need to consider the bias of algorithms and the potential for systemic discrimination. In 2016, ProPublica, for instance revealed that Facebook was enabling discriminatory targeted advertising, allowing advertisers to target only "white households." At the time, allegedly Facebook replied that they were going to address the issue. Yet in 2017, ProPublica carried out a further investigation to see whether measures had been taken. To do so, ProPublica bought dozens of rental housing ads on Facebook, but asked that they *not be* shown to certain categories of users, such as African Americans, mothers of high school kids, people interested in wheelchair ramps, Jews, expats from Argentina, and Spanish speakers (Angwin et al. 2017).

In accordance with the Federal Fair Housing Act in the US, it is illegal to publish any advertisement with respect to the sale or rental of a house that indicates any preference, limitation, or discrimination based on race, color, religion, sex, handicap, familial status, or national origin. Violators can face tens of thousands of dollars in fines. ProPublica, however, found out that every single ad was approved by Facebook immediately, and that the only ad that took longer to be approved—but was then approved—was one that sought to exclude potential renters interested in Islam, Sunni Islam, and Shia Islam (Angwin et al. 2017). In March 2019 following the ad scandal, Facebook had to implement measures to avoid such forms of racial and ethnic discrimination, yet the debate is still going on at the time of writing especially in relation to other sensitive data such as gender and politics (Gillum and Tobin 2019).

The Facebook ad scandal is just the most public example so far that highlights how sensitive data profiling can impact real life situations. Yet we have to realize that sensitive data is largely inferred through household profiling and home life data. Individuals have historically been profiled on the basis of the families and the social groups they belong to. Yet, today these classifications are made possible through algorithmic decisions that are difficult to understand or to challenge. The fact that we are allowing data technologies to profile households can have serious implications for children's rights and society as a whole. It not only reproduces existing inequalities and stalls social mobility (Eubanks 2018), but it can also impact children's rights to self-definition and nondiscrimination.

CONCLUSION

Home automation, with its complex data-environments and data flows, is radically transforming the ways in which we should think and protect children's privacy. By introducing the concept of home life data, the aim of this chapter was to argue that it is of central importance that we appreciate that companies are harnessing great amounts of data from the home that includes not only personally identifiable data but also data about the family socioeconomic context, values, and everyday behaviors. This data is used to profile households for targeted advertising. The implications of these processes of household profiling can be immense, they can lead to discriminatory behavior, and they can be used to jump to conclusions about families and children on the basis of messy and inaccurate data. Yet, the question about household profiling and the aggregation of adult and child personal data is crucial, not only when we think about home technologies but also for our understanding of social media data and its impact on children's lives.

6

SOCIAL MEDIA DATA: CHILDREN'S SOCIAL MEDIA PRESENCE AND PROFILING

One day in the summer of 2018 I sat in Scott's living room in Los Angeles for an interview. I asked him whether he could imagine how his children would be profiled on the basis of the images that he posted on social media. His answer shocked me:

S: Oh, they probably would be profiled as very adventurous, because we photographed them in all these places all over the world. Kind of corny because we photograph them in different costumes, and things like this. … I think it would be clear that they would be profiled as being Jewish, which is a bit scary.

V: How?

S: Well we have definitely taken pictures at events thrown by organizations that identify as Jewish, and those people have also reposted their picture, which is definitely very scary for me. I was a very active Twitter user up until probably mid-2016. During the election, an article came out about Steve Bannon's divorce proceedings. The article mentioned that Bannon did not want his children to go to a private school, here in Los Angeles, because "there were too many Jews." I knew what school that was, my girls were going to go there. But then we decided against it. So I posted a photo of the girls on Twitter and commented that I found his comment "pretty upsetting" because my "two-little Jewish girls, would probably go to that school." I got all these crazy hate messages and replies, saying all this crazy anti-Semitic stuff and people doxing my address and pictures of the front of my house. So after that I stopped posting on Twitter.

Before interviewing Scott, I had listened to a lot of parents talking about the impacts of their social media "sharing," but I had never encountered such a shocking and vivid representation of its risks. For one single Tweet—

written in anger—Scott's daughters had not only been exposed to anti-Semitic online abuse, but they were placed at risk through the publication of their home address. Of course, when Scott posted the photo of his daughters, it was 2016, a time when the discussion in mainstream media on children's online privacy and social media was just starting. When he posted the picture, Scott did not reflect on the risks. When he realized what he had done, he was already "a post too late."

For how shocking Scott's story was to me, it was clearly an exception. Yet I decided to start this chapter with his story because this is the type of story that defines contemporary debates that focus on the risks of so-called sharenting. The term *sharenting* was first introduced by reporters at the *Wall Street Journal* to describe parents who overshare information of their children's lives on social media. In the media we have seen a plurality of articles dealing with the topic. As this chapter will show, however, I do not find the term *sharenting* useful or enlightening in describing the impacts of social media data on children or the complexity of children's social media presence. This chapter will explore this complexity by mapping the everyday negotiations and conflicts that define social media data flows in family life. The aim is to argue that we need to move beyond the question about sharenting and we need to start looking at what data is actually being collected by social media companies and how they are profiling children.

BEYOND SHARENTING? CHALLENGING IDEAS OF BLAME AND IMPACT

In 2016, the news broke that an 18-year-old Austrian girl was suing her parents for posting her photos on social media. The news story appeared on different newspapers in the UK and the US, such as *The Independent* (Khan 2016) and *USA Today* (May 2016). The articles claimed that the 18-year-old girl was accusing her parents of having "no shame and no limit" in sharing more than five hundred photos of her childhood online. The articles also claimed that this was the very first case of a child suing parents for the practice of sharenting. A few days later, the news story was debunked by a reporter at the *Deutsche Weller*, who found out not only that the case was not registered in any Austrian court, but also that the lawyer who had been identified in the media as the lawyer of the 18-year-old girl had spoken about the case only in hypothetical terms (Perez 2016).

What made this case particularly interesting is not that it was yet another example of fake news, which was reported also by established media organizations, but that it spoke directly to the moral panic and the narrative of blame that is often associated with the very notion of sharenting. Of course, the social media practices of parents and others can build specific narratives of children that could impact their privacy or expose them to risks. Yet we cannot take for granted specific impacts and risks of social media posting, because we cannot predict them now.

I reached this conclusion when, in spring 2017, I met Mina. That day, I took the bus down Brockley Road in south London. It was a rainy day and I passed the coffee shops, the old cemetery in front of the ballroom, amazed about the ways in which that neighborhood had been gentrified. I continued my journey all the way to Catford, a working-class, low-income neighborhood in the heart of south-east London. I looked at my phone in search of the address and made my way following the blue line on Google Maps to a council estate. It was raining, cold, and almost dark. Mina opened the door, and a smell of delicious food came from her kitchen. Her 5-month-old baby was in her arms and she was impressed by how far into my pregnancy I was. We started to talk about feedings and naps; we talked about my eldest, and she asked questions about the interview that we were about to start.

M: The interview is about social media right? Because I post a lot. A lot. I post all the time. Do you post?

V: No. I don't.

M: Not even for your first child? Not even for your pregnancy?

V: No, my husband and I kind of made that decision when P was born.

M: Aren't you afraid that she is going to resent you for not having photos on social media?

V: Well, maybe. The truth is that we don't know.

When Mina asked me that question, I immediately realized that we know very little of the many different impacts of social media, especially because these impacts are not determined by the world that we live in now, but by the techno-social environments of the future. When Mina asked me that question, I realized that I had been biased and narrow-minded when I tried to predict the possible impacts of social media and protect my children. Paul and I chose not to share information of our children (or our personal lives) on social media platforms because we thought we were protecting our children's right to privacy in public.

After Mina asked me that question, I started to wonder whether our choice of not sharing information on social media would actually impact my children in negative ways in the future. What if Mina was right that my daughters might come to resent me for not posting about them? Or even worse, what if they are going to be refused insurance or a job position simply because there is not enough social media data for future companies or AI to assess them with?

Of course, I am speculating here. Yet we are all speculating when we try to predict how social media data will actually impact children in the future. Therefore, rather than blaming parents about their social media habits or focusing on risks and impacts that we cannot predict, it might be important to actually analyze the complexity of children's social media presence and social media data.

CHILDREN AND THE IDENTITY OF OTHERS

In the last few years, much attention has been placed on children's social media presence. In 2010, the internet security company AVG surveyed mothers in the US, Canada, UK, France, Germany, Italy, Spain, Australia, New Zealand, and Japan, and found that 92 percent of children had an online presence by the age of 2 (Business Wire 2010). A 2015 study by Parentzone .com, on behalf of Nominet, found that the average parent posts allegedly around 1,500 pictures of their child online before the fifth birthday (Rose 2015). Such studies have been accompanied by a plethora of newspaper articles on the dangers, risks, and implications of parents' social media habits.

What these debates show is that one of the greatest complexities of social media data is that this data is defined by the coming together of multiple different identities under a unique digital self (Blum-Ross and Livingstone 2017). Parents often share information about their children online, not only to show what great kids they have but also as a form of self-representation and as a way to build their own identities as parents. One day, during the Child | Data | Citizen project, for instance, I interviewed Bonnie in Los Angeles. Bonnie arrived late with her 6-month-old baby in his stroller, wearing a "don't mess with me" bodysuit. I laughed, when I read it, and Bonnie was very pleased that I noticed. Bonnie had been working as a waitress in bars all her life. Her boyfriend who was from the UK was a bar manager.

They had just moved to LA from the East Coast. At the time of the interview, she was in her early thirties.

Bonnie was not a prolific sharer. Yet she did post regularly about her baby, to "keep in touch with her friends back home" and "to keep her boyfriend's family updated." This is how she described her social media use during the interview and the overlapping of her and her baby's identity:

B: I didn't post on Facebook that I was pregnant. So, I was in the hospital and said: 'surprise!! It isn't a burrito.' I got around 100 likes. I never got 100 likes before. You know, people don't actually like me but they fricking love babies. … I post photos of him doing something fun with the cat, or him in the bathtub with his dad, or me drinking a beer with him in my arms. They have to be worth it; I don't want to lose friends because I have a baby."

V: Do you think you are building a digital narrative when you do that?"

B: Yes! I am. I am building a narrative that says: 'Look she is still fun even if she has kids! We can still relate to her; we don't hate her.' I am really concerned about what people think. I don't want my friends not to like me anymore because I have a baby."

V: Right but this is about your digital narrative, what about the digital narrative that you are building for him?"

B: [Laughs] Oh yes the narrative I am building for him is: 'Look at how my mum used to fuck with me.' No, I am joking, we try to be funny, to make him funny, but we try to be not too embarrassing. [Laughs again.]

Bonnie was using social media to negotiate with her new identity as a mother and with her own fears about that identity. She wanted to come across as fun, even if she had a baby. She was also using social media to share the life of her child with the family and friends who lived far away. Bonnie was aware that through her social media posting she was not only building a narrative about herself, but she was also constructing a narrative for her baby. This understanding, in its simplicity, is what defines the complexity of social media presence: each post, because relationships are involved, tells the story for different lives.

So-called parent bloggers are certainly the most extreme example of this and have become key figures for thinking and arguing about children's right to privacy. On January 3, 2019, for example, the mother blogger Christie

Tate, wrote an opinion piece for the *Washington Post*. In her article, Tate tells the readers that when her fourth-grade daughter found out about her parenting blog, she was very angry and "asked her to take the essays and photos that were published off the internet." Tate explains to her audience that even if she understood that she was impacting her child's right to privacy, she was not going to be able to stop "exploring her motherhood" in her writing (Tate 2019). When I read her article, I immediately wondered how my children will respond to my book.

The article received more than seven hundred very negative comments, in which users expressed all their outrage against Tate for putting her needs before her children's privacy. After the backlash, Tate hasn't been Tweeting publicly and her blog is now private. Even if Tate's example is certainly one of the most extreme, it clearly shows that when we think about social media data there is a tension between the rights of the parent to self-express and the right to privacy of the child, and this is precisely because there is an overlapping of digital identities.

Through their everyday posts, parents build a specific public persona for their children. In 2016, for instance, I carried out a research on parent activists and realized that they constructed their children as "political subjects" (Barassi 2017). Over and over again I came across posts of children at demonstrations holding banners and flags; children playing together and being described as "plotting the next revolution"; children with signs openly criticizing the current government or supporting specific political campaigns. Hence, parents' social media practices were somehow building the political identity of the children through processes of narrative construction.

From a data privacy point of view, that each post talks about the overlapping of the child and the parent identities is problematic. In fact, the construction of parents' digital self depends on making the lives of their children public (Blum-Ross and Livingstone 2017). Therefore, parents' right to self-expression and self-representation is seen as directly impacting their children's right to privacy. These questions have been explored by legal experts such as Steinberg (2016) and Bessant (2017), for instance, who considered the legal implications of sharenting in the UK and US and explored how children can protect their agency and seek legal action if they wanted. As legal scholars, journalists, and researchers debate in public about the social and legal ramifications of children's social media presence, families are actively negotiating with the complexity of social media data in their everyday life.

CHILDREN'S CONSENT AND FAMILY TENSIONS

One day in 2017, I was sitting in Los Angeles with Zoe, who was married to Mike and who I introduced in chapter 1. Zoe had a 5-year-old daughter and a 12-year-old stepson and was a sharent. She liked sharing information of her children online. During the interview, Zoe told me that she had posted the first picture of C—her five-year-old—"as soon as she was born, right from the delivery room." After her birth, she posted news feeds of her daughter on Facebook at least twice a week. This was important to her because her family lived on the East Coast and Facebook played a key role in connecting them. Zoe was not concerned about the privacy implications of her posts, because she only "shared good stuff" and, if someone were to look at what she posted "they would figure out stuff about C, but only good details" and then she added: "They would know that her mom really loved this kid and that she is a pretty great kid." Zoe told me that C knew about Facebook and that many times she would actively ask her mother to post images or content.

Zoe's point about posting positive images and about C's involvement in her social media posting is key to parent–child relationships when it comes to social media use. Moser et al. (2017) from the University of Michigan published a study based on a survey of 331 parent–child pairs, which examined parents and children's preferences about what parents should share on social media. In contrast to much smaller studies (Lipu and Siibak 2019) or articles based on a handful of experiences (Lorenz 2019), which stress children's frustrations or shock about seeing their photos online, Moser et al. (2017) demonstrate that children were OK with their parents posting cute or fun family pictures or pictures that made them look good; they perceived these photos as flattering. Yet they were against parents posting embarrassing photos, ugly pictures, or intimate photos that show them for what they really are like at home.

I find Moser et al.'s (2017) work particularly fascinating, because it clearly shows that the practice of sharenting is being negotiated piece by piece by parents and children. Families are actively negotiating with this issue of children's consent and are strategically thinking about their own social media practices. One couple of parents I worked with, for instance, would never post images of their children online without asking permission; another couple would never post at all. Among those parents that did post often, the issue of children's consent was certainly becoming an important aspect of their digital practices.

One afternoon, for instance, I had the pleasure to sit down for an interview in a crowded Starbucks cafe near Angel Tube station in London with Caty, who worked as a civil servant and had a son who was 6 years old. I had randomly met Caty ten years before through a mutual interest, and we became Facebook friends then. We had not seen each other since 2007, but I witnessed the birth of her child and followed him growing up on Facebook. It is for this reason that I asked her to participate in the project.

During the interview I asked Caty to explain what social media posting meant to her, and she replied:

C: For me [Facebook] is like a visual diary. I use it because it enables me to remember things that made me laugh, details of our everyday lives, or things that made me angry or made me think. It is about my emotional response. I know M has such a big online presence. But it is really because my mom and dad live miles away and for the rest of the family. It is a very, very easy way to share stories and see M's personality develop, and it is hilarious.

V: You also have a very fun way of telling stories of M on Facebook.

C: Yes, it's true [laughs]. I think it's a nice way to see how our life is evolving and it's an interactive way for others to participate in it, so that he knows that the others were watching him growing up.

V: So, what does M say about you posting?

C: Well he has an idea of what Facebook is. He knows that we share stories and pictures and things that we like. We haven't had the "consent" talk yet, but it's in the list of things to do. You know if he tells me don't post this, I respect that. Sometimes he tells me that he doesn't want me to take a picture and I say fine; I won't do it.

The experience of parents like Caty reveals that families are actively thinking about what they are posting online and how they share information about their children. It also reveals that they are involving children in their social media practices and are critically thinking about the issue of consent. If we want to understand the impact of social media data on family life, however, we cannot only focus on parent–child relationships and on the issue of children's consent. We need to consider how, within families, social media data has become a contested terrain of meaning for families, which is subject to everyday negotiations and tensions among different family members.

Social media–related tensions and negotiations were a norm among partners or co-parents. Frank and Mandie, who lived in south London and had a 3-year-old girl, were definitely experiencing these tensions. Mandie did not agree that Frank posted so many images of T on Facebook, because she didn't really know who his friends were and, according to her "he had hundreds of friends." Yet for Frank, posting on social media was a way to keep in touch with his friends and family and to have them participate in his life. Also, in Marie's family there were tensions. Marie was from London, but she had moved to LA because her husband was American. They lived in Santa Monica with their 6-month-old baby. Marie liked to share, but she did not share much at all because her "husband did not allow it."

Social media–related tensions did not only involve co-parents but a plurality of other relationships in family life. Sometimes parents expressed their frustration against a relative or a friend who they identified as "the sharer." Other times they complained about their mother, friend, relative, school, extra-curriculum activity, and so forth, for posting pictures of their children. When we start looking at all these daily tensions, a complex map of children's data flows emerges, which goes well beyond what parents share or do not share. This map is shaped by the variety and plurality of social relationships that define everyday family life and is almost impossible to visualize.

SOCIAL MEDIA DATA AND THE IMPOSSIBILITY OF CONTROL

What I realized by talking to parents and trying to map the social media flows of their children was that these data flows systematically escaped their control. Katie, who I introduced in chapter 2, for instance, believed that she was being careful and limiting the data that she posted, but then admitted to tagging her fiancé and not knowing how many friends he had. Annie, who lived in south London with her 4-year-old son, her 4-month old baby, and her husband Guy, and who had just moved to the UK from the Caribbean was very protective of her children's privacy on social media. Yet during the interview she explained that in the day-to-day life, it was extraordinarily difficult for her to control her children's data flows:

A: When my daughter was born I announced that she was born by posting one picture on Facebook. It was a picture of her and my husband that had their back to the camera, so no facial features.

> That's all. We don't post at all, and nothing for this little guy [she looks down at the baby who was breastfeeding]. People around us feel very different about this issue, and it has become awkward for them not to post about our children. Also for us, I feel that it has become so difficult to protect their digital image and control it on a day-to-day basis. At nursery for instance we told them not to take photos. There have been tensions especially during the Christmas Pantomimes, because we made it clear that we didn't want anyone in the audience to take photos and other parents were upset because they wanted to take photos of their children.

For those parents, like Annie, who have deliberately chosen not to share information on social media, it has become incredibly difficult to control the social media data flows of their children. They have to constantly engage in daily conflicts with others who instead believe in the importance of social media data.

For how much I tried to keep my daughters off social media, there have been multiple situations when I just accepted that other people took pictures of them and may have shared them on social media. One day in particular, I took P to a park, where she met a little friend and they started to play in a sandpit. The grandfather of the little girl who was playing with my daughter, and whom I never met, looked very proud that his granddaughter—who was much younger than mine—was socializing. He started taking pictures. I was sitting on the bench and wondered whether he posted those pictures on Facebook or other social media. I also wondered whether through facial recognition it was possible to identify my daughter and locate her at that given time in that given place. I felt uncomfortable, but I didn't have the heart to confront him on the images that he had just taken.

Once we understand that in everyday family life, the social media data flows of children are extraordinarily complex and almost impossible to control, the following questions arise spontaneously: What type of data are we producing on social media? How are companies collecting, sharing, and using it in the age of surveillance capitalism? What is happening to the data of children?

WHAT DATA IS COLLECTED FROM SOCIAL MEDIA?

In order to understand social media data, we need at first to take a quick step back in history and appreciate how the business model of Web 2.0 technologies was developed. In 2004, at the first Web 2.0 Conference, Tim O'Reilly (2005) explained that the early 2000s had seen the development of a different type of web, the Web 2.0, which harnessed "the collective intelligence of crowds to create value" (O'Reilly 2005, para. 25). The new web, in contrast to Web 1.0, was no longer based on a network of hypertexts, but was defined by a new "architecture of participation" (O'Reilly 2005, para. 24).

The defining features of Web 2.0 included rich user experience (easier to navigate), user participation (users could interact and engage with the content), dynamic content (content would continue to change, and old content would be accessible), metadata (content would be defined also by the context), valid markup (introduction of new web standards), and scalability (reviews) (Best 2006). The new business model enabled the development of different social media platforms that were released in those years: Facebook was released in 2004, YouTube in 2005 (and bought by Google in 2006), Twitter in 2006, and Snapchat in 2011.

From the very early days of social media platforms it became clear to some that these technologies were in fact tools for the surveillance, storage, and exploitation of large amounts of personal data (Andrejevic 2004; Terranova 2004; Van Dijck and Nieborg 2009). In the last fifteen years, social media companies have collected, shared, and processed unimaginable amounts of personal data to make profit. We do not know exactly how this data is being used to profile individuals. As Pasquale (2016) has shown, these companies rely on trade secrecy, which makes it impossible to really understand what happens to our personal data. Yet from a platform analysis of different social media companies, we can start mapping "what companies say they do" with our data and with the data of our children.

During the Child | Data | Citizen project, I carried out a platform analysis of four different social media companies: Facebook, YouTube (Google), Snapchat, and Twitter. The analysis, as explained in the introduction, consisted in an exploration of the promotional cultures, business models, and data policies of the different platforms. It also consisted in researching and following the different news that involved patent requests as well as privacy scandals. The platform analysis was largely qualitative and ethnographically informed in the sense that I found myself analyzing these platforms simultaneously as a researcher and as a concerned parent who wanted to find out how my children's social media data (and my own) were being collected and used.

During the research, I came to the simplified (and very public) understanding that broadly speaking there are five different types of data flows that companies admit gathering in their data policies:

1. REGISTRATION LOG-IN DETAILS: Details include name, date of birth, and email to set up the account; sometimes people register other important details such as workplace, education, and so forth.
2. ACTIVITY: Social media companies collect a lot of data in relation to "what people do" when they use their services. Of course, activity data varies from social media platform to social media platform. Yet broadly speaking activity data includes the following: "How you interact with content"; "Voice and audio information when you use audio features"; "Purchase activity"; "People with whom you communicate or share content"; and "Search activity."
3. CONTENT: Social media companies collect a lot of data on the content we produce. Google, for example, collects different kinds of content: "Emails you write and receive, photos and videos you save, docs and spreadsheets you create, and comments you make on YouTube videos" (Google LLC 2017). All the four companies shared very similar terms of services when it came to the issue of content ownership. They all made sure that they clarified that "Users retained data ownership and intellectual property rights" of the content they produced. Yet by signing off their terms of service, the user also agreed to grant the company (and their "affiliates," "partners," or "those who they work with") "A worldwide, royalty-free, sublicensable, and transferable license to host, store, use, display, reproduce, modify, adapt, edit, publish, and distribute that content" (Snap Inc. 2019), yet similar wording is used by all the four social media companies, including Google (2017) (figure 6.1). Hence even if the user retains ownership of content, social media companies can do whatever they like with it.
4. DEVICE DATA: Social media companies gather much personal information from devices. This is personally identifiable data, such as "IP address, unique identifiers, device IDs, and other identifiers, such as from games, apps or accounts you use, and *Family Device IDs*, (which again suggests that companies are actively profiling family units, italics added). They also gather "Information you allow them to receive through device settings you turn on, such as access to your GPS location, camera or photos" and other data such as "Battery level, … browser type, app and file names; cookie data etc." Companies are also really trying to collect more

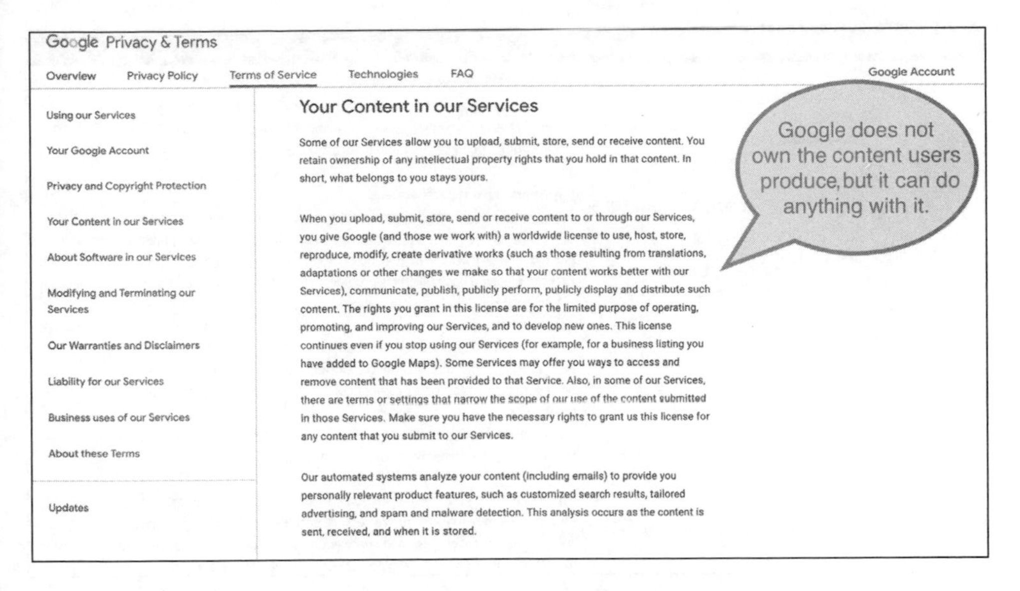

Google Privacy & Terms

Overview | Privacy Policy | Terms of Service | Technologies | FAQ | Google Account

Using our Services

Your Google Account

Privacy and Copyright Protection

Your Content in our Services

About Software in our Services

Modifying and Terminating our Services

Our Warranties and Disclaimers

Liability for our Services

Business uses of our Services

About these Terms

Updates

Your Content in our Services

Some of our Services allow you to upload, submit, store, send or receive content. You retain ownership of any intellectual property rights that you hold in that content. In short, what belongs to you stays yours.

When you upload, submit, store, send or receive content to or through our Services, you give Google (and those we work with) a worldwide license to use, host, store, reproduce, modify, create derivative works (such as those resulting from translations, adaptations or other changes we make so that your content works better with our Services), communicate, publish, publicly perform, publicly display and distribute such content. The rights you grant in this license are for the limited purpose of operating, promoting, and improving our Services, and to develop new ones. This license continues even if you stop using our Services (for example, for a business listing you have added to Google Maps). Some Services may offer you ways to access and remove content that has been provided to that Service. Also, in some of our Services, there are terms or settings that narrow the scope of our use of the content submitted in those Services. Make sure you have the necessary rights to grant us this license for any content that you submit to our Services.

Our automated systems analyze your content (including emails) to provide you personally relevant product features, such as customized search results, tailored advertising, and spam and malware detection. This analysis occurs as the content is sent, received, and when it is stored.

Figure 6.1
Google terms and conditions (2019).

specific behavioral data, which basically means "Information about operations and behaviors performed on the device, such as whether a window is foregrounded or backgrounded, or mouse movements" (Facebook 2018a).

5. DATA BROKERING OFF-LINE: All companies admitted collecting data on social media users off-line through third-party partners. *Even in the case in which users do not have an account* (figure 6.2). Facebook for instance lets its users know that "Partners provide information about your activities off Facebook—including information about your device, websites you visit, purchases you make, the ads you see, and how you use their services—whether or not you have a Facebook account or are logged into Facebook … We also receive information about your online and off-line actions and purchases from third-party data providers who have the rights to provide us with your information" (Facebook 2018a). Also Twitter mentions that they receive information when you view content on or otherwise interact with their services, even if you have not created an account (Twitter 2019a).

When I tried to break down all the different data that the companies collected, my intention was to be brief, concise, and to the point. I wanted to make data policies palatable for those parents who, as we have seen in chapter 2,

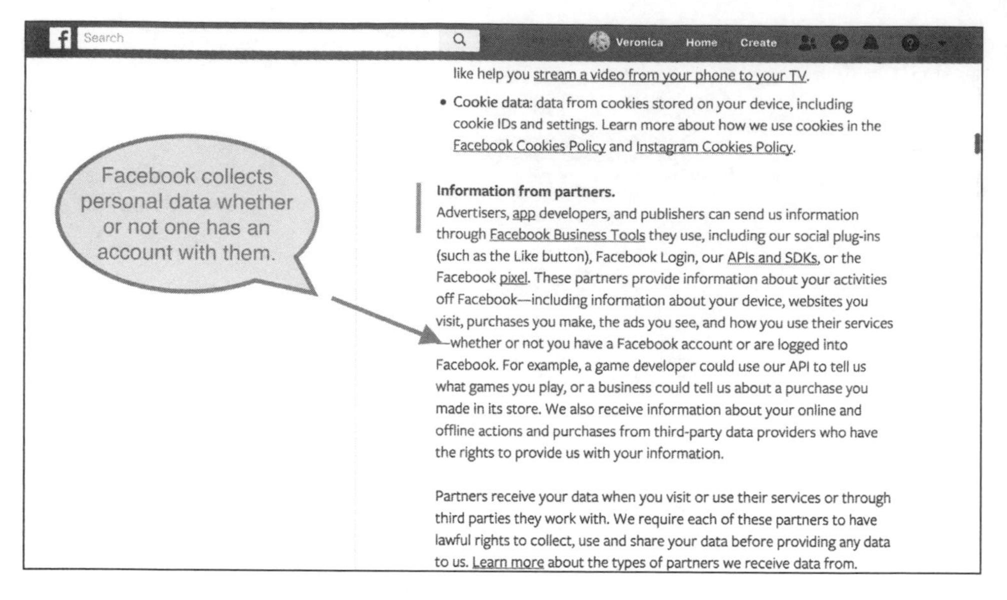

like help you stream a video from your phone to your TV.

- Cookie data: data from cookies stored on your device, including cookie IDs and settings. Learn more about how we use cookies in the Facebook Cookies Policy and Instagram Cookies Policy.

Information from partners.
Advertisers, app developers, and publishers can send us information through Facebook Business Tools they use, including our social plug-ins (such as the Like button), Facebook Login, our APIs and SDKs, or the Facebook pixel. These partners provide information about your activities off Facebook—including information about your device, websites you visit, purchases you make, the ads you see, and how you use their services —whether or not you have a Facebook account or are logged into Facebook. For example, a game developer could use our API to tell us what games you play, or a business could tell us about a purchase you made in its store. We also receive information about your online and offline actions and purchases from third-party data providers who have the rights to provide us with your information.

Partners receive your data when you visit or use their services or through third parties they work with. We require each of these partners to have lawful rights to collect, use and share your data before providing any data to us. Learn more about the types of partners we receive data from.

Figure 6.2
Facebook data policy (https://www.facebook.com/privacy/explanation).

would not really have the time to read all the different data policies. Yet I realize that the preceding information is still pretty overwhelming. This is because the data that social media companies collect is enormous.

When we think about social media data, however, we should not only think about the incredible amount of data that is produced, collected, and archived, but we also need to think about the multiple ways in which social media companies are creating a single system in which all the data traces can be mapped to a single profile ID. What I realized during my research is that parents are aware of the complexity of social media data and how easily it would be to profile someone on the basis of their social media presence. For instance I interviewed Linda, who lived Los Angeles with her two children aged 5 and 14 years. I asked her to describe the data flows of her family:

L: We have social media, and we produce quite a bit of information, I hope it doesn't affect my children's future, because I can't help it.

V: What type of information do you post?

L: Loads, also I play games on Facebook so they collect data that way too, and they figure out what type of foods I like, what ethnicity, the kind of person I am because of my values, they can also infer politics from that, they have a sense of my education, of where I worked, location, if someone is really studying me they might be able to figure out what time I sleep, what time I wake up, and infer that type of information. They can probably tell what type of entertainment products I like, they could infer income, and that I have two kids in private schools. They can tell a lot about us by looking at friends.

V: So let me recap, only from your social media you think people could be able to come to conclusions about class, education, ethnicity, politics, food choices, daily routines, entertainment choices . . .

L: Yes, precisely, they can build like another me.

When we think about social media data, therefore, we need to realize how this data is being used to create unique ID profiles, profiles that tell stories about individuals; that—as Linda mentioned—can build "another me." This is exactly what companies are trying to do. According to one of the latest books on data harnessing (Chavez et al. 2018), people create large amounts of personal data from a plurality of devices and platforms. What successful companies need to do is to invest in "people data" and thus map "distinctive data signatures" to single persons (Chavez et al. 2018, 26). I like reading business books because I believe that they provide us with insights into how businesses are thinking. At present, what really seems to be important to companies is to develop the means to integrate vast amounts of data, no longer in an aggregated form, but in a granular way—a way that enables companies to really understand individuals' profiles.

Big Tech companies are gathering large amounts of data of children precisely because they want to build in-depth, unique ID profiles of individuals across a lifetime. It is for this reason that companies are using biometric technologies, which can recognize individuals through their face or voice, and are tapping into multiple typologies of data from health to education, from home life data to social media data. Hence, we need to ask ourselves, How are social media companies collecting children's data? What are the implications of these forms of data collection?

CHILDREN'S SOCIAL MEDIA DATA

All the major social media companies comply with COPPA (Child Online Protection Act) and make it clear that their services are not available to children under the age of 13. This implies that, legally the four social media platforms that I analyzed are not designed and targeted at children under the age of 13, who cannot open a profile. For these reasons, when one reads the privacy policy of social media technologies there is very little mention of children. Yet it is clear that companies are trying *directly and indirectly* to harness the data of children beneath 13 years of age.

Companies are directly trying to harness children's data through the design of technologies that are explicitly targeted at them. In February 2015, YouTube/Google launched the YouTube Kids app. In June 2018, Facebook launched the Messenger Kids app. Both apps, because they are designed for children beneath the age of 13, have specific data policies that explain the degree of children's data collection and processing.

From the YouTube Kids policy, it is clear that Google is limiting the degree of data collection in the sense that the company focuses on the collection of device information, log information, and unique application numbers (similar to cookies used to collect and store information about an app or device, such as preferred language, watch and search history, and other settings). There is no direct mention to the collection of content, and Google claims not to be collecting name, email, or personal address from children.

In the case of Facebook Messenger Kids, the data collection is far more extensive, because they collect almost all the aforementioned categories: personal name and other personal details upon registration, activity (including networks and contacts), content, and device information. To give you an example, in their privacy notice they explicitly state the following:

> We collect the content and information your child sends and receives on Messenger Kids, such as the content of messages (including text, audio and video), stickers, gifs, photos or videos they send, camera effects they use, and their score in a game they play with a friend" and also we "collect information about how your child uses Messenger Kids, such as who they engage with, what features they use, and how long and in what ways they engage with different features on the app . . . the people your child connects with on Messenger Kids and how they interact with them, such as the people they communicate. (Facebook 2017)

If one reads the extent of their data collection, it is not surprising that in the last few years both technologies have been heavily criticized by lawmakers and privacy campaign groups. Examples such as Facebook Messenger Kids and YouTube Kids[1] reveal that social media companies are explicitly targeting children under the age of 13.

Although it is clear that social media companies are directly harnessing the data of children before the age of 13, we also need to realize that social media companies are gathering large amounts of children's data (under the age of 13) in indirect ways through adult profiles. All the pictures of children shared on Facebook by parents and friends, all the cute videos on YouTube, the hashtags on Twitter, and the snaps on Snapchat are collected according to terms that generally apply to the adult profile. In this way the data of children is integrated with the data of adults. Yet what we need to understand is that these companies have the means and technologies, or at least are developing the means and technologies, to track entire families (or households) and individually identify its members (including children).

In November 2018, for instance, as mentioned in chapter 5, Facebook filed a patent application for "Predicting Household Demographics Based on Image Data" (Bullock et al. 2018). The technology, which relies on facial recognition, will enable Facebook to profile "photos posted by the user" and "photos posted by other users socially connected with the user." They also will be able to analyze textual data (e.g., captions such as "thankful for my family") to "build more information about the user and his/her household in the online system, and provide improved and targeted content delivery to the user and the user's household" (Bullock et al. 2018). The question that the patent request does not seem to be answering however is the question about children's data. What will happen to children's data? Will Facebook identify children on the basis of the photos and content scanned? Will they assign a number (known as template) to a child's facial recognition?

SOCIAL MEDIA DATA AND AI PROFILING

When we break down exactly what type of data and how much data is being collected by social media companies, and the steps that they are taking to profile households, we certainly start to grasp the complexity of social media data. Yet focusing on the type of data companies collect is really only just scraping the surface, because the real question that we need to ask is one not about data collection but about data inference. Today social media data is constantly used for *personality profiling* by a plurality of social actors. The school headmaster, the employer, the insurer, and the potential date constantly check the social media profiles of others in order to reach conclusions about their behavioral or psychological characteristics.

Although this is a shared anthropological phenomenon, what is also happening is that AI technologies are increasingly being used to scrape social media data and make automated and data-driven decisions about individuals. These technologies and practices can lead to serious data harm for families and children. During the Child | Data | Citizen project, I was particularly fascinated by the case of Predictim, a new babysitting recruitment service that vowed to support parents with "making the right decision about who to employ" through background checks and personality assessments based on social media data (Harwell 2018). The company relied on AI technologies to generate a personality assessment of potential babysitters by analyzing social media data and focusing on a person's speech and facial expression. The potential candidates would be then scored as "at high risk" or "low risk" for abuse, bullying, disrespectful behaviors, and other individual characteristics. The candidates who refused to be assessed by the service were told that they would lose competitive advantage. The parents who used the service did not have to share the results of the assessment with potential candidates.

The story (Harwell 2018) in the *Washington Post* was shared multiple times on social media. What made the case particularly fascinating was that the service was designed to analyze teenagers and low-income candidates misuse of social media to rank them and judge them as potential abusers. When I read the story about Predictim I thought that it was a perfect example of how social media data coupled with nontransparent AI decision making can impact a person's life. As the story unfolded in the news, parents and prospective babysitters who used the service voiced their concerns about the inaccuracy and bias of the service.

In November 2018, Twitter, Facebook, and Instagram blocked the AI from their platforms, arguing that the technology violated their privacy policy. In December 2018, Brian Merchant, a writer at Gizmodo, wrote about how he used the technology to scan a handful of people that he would trust with his son and was surprised to find out that his son's babysitter, who had always been reliable and great with his son, was ranked worse than one of his friends, who "routinely spewed vulgarities on social media" (Merchant 2018).

Merchant publicly questioned Predictim over these misconstrued rankings by stating that the only difference between his babysitter and his friend was that she was black and he was white. The company's Chief Technology

Officer (CTO), Joel Simonoff, assured Merchant that he could guarantee 100 percent that there was no bias involved in the analysis (Merchant 2018), which as we will see in future chapters, is either an insincere statement or completely naïve. Following the backlash, in December 2018, Predictim's website announced that the company would indefinitely postpone the full launch of their services.

The case of Predictim shows what happens when journalists, researchers, and members of the public come together to question and scrutinize the inaccurate and biased assessments and scores of AI technologies, especially when these assessments are made on the basis of social media data. Yet the problem faced today is that social media data is being used to make data-driven decisions about who we are and to talk about us in public in a plurality of settings. For children whose lives have been shared on social media from the day they were born and who have not been in control of the data and narratives that have been constructed about them, social media data—as we shall see in the next chapter—may prevent them from becoming their own persons, because it can impact their right to self-definition, moral autonomy, and nondiscrimination.

CONCLUSION

Rather than focusing on parents' sharing practices, we need to fully appreciate the complexity of social media data in family life and stop believing that the danger of social media data is represented by parents who share images of children on social media. The real danger of social media data is defined by the fact that more and more businesses and organizations—from insurers and government officials to recruiters—use that image for data-driven decision making. This is the main risk of social media data, not the data itself, not that the parent chooses to share images of the child, but that as a society we started to believe that we can draw conclusions about people on the basis of their social media data.

The other real danger of social media data is that Big Tech companies like Google and Facebook have the means and technologies to build unique ID profiles to track children through time and across different spheres of social life, with little transparency and accountability. The tracking of individuals across a lifetime is what defines their business model. It is for this reason that they are harnessing health data, home life data, education data, and multiple other forms of data that children and families produce in their everyday life. As Paul, my husband, once told me: "The real danger is what these companies are taking from us, not what we share." As we shall see in the next chapters, I think he is right, because what companies are taking away from us (and from our children) is our right to self-construct and self-define in public.

7

DATAFIED CITIZENS: CHILDREN'S PRIVACY AND THEIR PUBLIC SELF

During the 2016 US presidential elections, I met Jen in west Los Angeles for coffee and an interview. As soon as she sat down and before the interview started, she told me that the previous evening her 6-year-old daughter had an emotional meltdown at the prospect that Donald Trump was going to be the next president of the United States. Jen is a Democrat, and she explained to me that in the run-up to the elections, she had discussed the presidential campaign with her children, and her eldest had become particularly involved and interested in the process.

When she told her daughter that Trump was probably going to win, her daughter started to cry and shout, "I don't want Trump to win, I don't want him." Jen felt very proud about her daughter's sudden political interest and found it incredibly amusing. So, she took her phone out and started filming. After she filmed her, Jen shared her video with her husband Carlos, who I introduced at the beginning of the book, and told him that she was thinking about putting it on Facebook. Carlos dissuaded her from doing it, because he was concerned that it would create an important political trace, a trace that could impact their daughter's future.

Jen and Carlos were not the only parents I interviewed who were actively questioning how data traces would impact their children's future. During the research many parents were interested in knowing my opinion on the matter. Are the data traces of children really going to impact their future? I have also been approached by journalists who asked me these questions especially with reference to social media data and sharenting. I always replied that I did not know. I believe it is impossible for us to predict the future, to fully understand the social and political implications of children's data traces. As mentioned in chapter 6, we really do not know how this data could impact the generation to come. Will social media data be available in the future? Maybe not. How will it be used and processed? We do not know. Will a lack of a social media presence impact a child's future life like a bad credit score? It might.

Even if it is impossible to foresee the future, what we know for sure is that in our current data-driven economies, *data traces are speaking for and about individuals* in ways that were not possible before, and these data traces are being collected, processed, and used to build narratives about people from before they are born. Rather than trying to predict how data traces will be impacting our children's future, I believe that we should take a step back and consider how surveillance capitalism is transforming the very notion of public self and how this is impacting a child's right to moral autonomy and contextual integrity (Nissenbaum 2010). This chapter will bring together different theories on personhood, self-representation, and digital citizenship. If we want to understand how surveillance capitalism is impacting our democratic life we need to critically explore how data traces challenge citizens' right to self-representation, how they *speak for and about us in public,* and how they are turning us into datafied citizens from before birth.

DATA TRACES AND THE PUBLIC LIFE OF THE CHILD

The day I met Maya she was sitting in a coffee shop of a library in Los Angeles with her 8-month-old baby in her arms, chatting to other mothers while scrolling through her phone. I approached them to see if they wanted to participate to my project, and I found myself a few days later knocking at her door. During the interview, Maya told me that she was a passionate sharent; she loved to post about her son and tell stories about his life. She posted daily on Facebook and Instagram and created a hashtag only for her son. Through the hashtag Maya reflected on the joys and challenges of same-sex parenting; on life as a new mother; and on fashion, entertainment, and politics. I asked her whether she thought that she was building a public image of her son on social media:

M: Oh yes because you always post the best picture and you don't really post what got you to that point. You don't post if he had an explosive poopy diaper or if you were up all night. But what you present is a smiley baby that looks super cute. I don't post ugly pictures; I don't show the bad stuff. … I just present this funny little guy. I want to tell everyone that we are having an amazing adventure and that life is great.

V: But what type of narrative about his life do you think that people would build if they looked at your social media accounts?

M: Obviously they would know that he was brought up by two moms. They would know what he ate as a child, so about his health; what activities he has done ... they would know about politics because we went to Hillary's rally and that was cool. We have a picture of him as we held him up and you can see his face and Hillary's face, and we posted it.

V: Wow, that is a lot of personal data and aren't you afraid that he could be profiled on the basis of that data?

M: Yeah, but he is not running for office ... and plus we believe in the stuff we publish so if people criticized us for this we wouldn't want to be associated to them.

V: But what about him? What happens if in the future he wants to choose his own values and not be associated to yours? Do you share the same values of your family?

M: Oh no, no. I was brought up as a Christian and my mom is Republican.

V: Oh right. So how would you feel if your mom published photos of you when you were little at a Republican rally?

M: Oh, that would be embarrassing. Thank God the internet was not around back then (laughs). I guess I would try and justify the photo and explain that it is not "me" now. But I guess we are pushing our values on him, and he does not have a say or choice. I guess he could choose later on in life. But ... oh man! ... What if he wants to be a Republican president and there is a baby photo of him with Hillary Clinton in the background? Wow, we don't think about that do we?

One of the main problems of children's data traces is that—as Maya's example shows—they can impact a child's ability to construct their public self and on their right to choose which values to be associated with and which not. In fact, the construction of parents' digital self on social media depends on making the lives of their children public, and hence building a narrative of their children's public self (Blum-Ross and Livingstone 2017). In addition to that, the problem with children's data traces goes much further than parents' narratives on social media. In everyday life it is not only parents that are building narratives about them, it is a plurality of other technical and human agents who produce, harness, and use children's data to profile them in one way or another.

When we think about all these data traces, we can realize that the difference between my daughters and me is that there's very little information about my childhood out there. There's no record of whether my mom smoked when she was pregnant, whether any members of my family were right or left wing, or of all the embarrassing things I did when I was a teenager. I could grow up to be my own person, to choose my own values and live by them. My children won't have that liberty. For the rest of their lives, they may be judged by the data points collected through their childhood. Data traces can build public narratives of children that can follow them across a lifetime and impact their ability to construct their public self.

The construction of one's own sense of public self is extraordinarily complex. It often begins from childhood as children are socialized into specific values by their parents, their educators, their friends. They learn how society works, internalize understandings, and start making conscious decisions about the type of values and beliefs that they want to follow. It is for this reason that the philosopher Hegel argued that political identity develops from within the family *and* within civil society as individuals develop the capacity to decide which inherited traits or qualities they want to endorse and which not (in Moland 2011, 37).

The construction of one's own sense of public self is thus an open-ended process of approximation and distancing from the values that we learn in our families, because we tend to measure these values up against other values and social experiences that we learn in the encounter with others (e.g., school, friends, extracurricular activities, to name a few). Unfortunately, there is not much research on the public life of the child and on how public self-construction begins from childhood (Nolas et al. 2016). Melissa Nolas, a sociologist at Goldsmiths, University of London, over the last five years has been working tirelessly on a project funded by the European Research Council, which studies how children construct their very sense of publicness and public self. Despite that research on children's sense of public self is scarce, for more than fifty years researchers have highlighted that individuals' civic values are shaped during childhood, and that children actively negotiate with the values offered to them by the family, school, and society as a whole (Liebes and Ribak 1992; Hess et al. 2005).

THE PUBLIC SELF AND OWNING ONE'S NARRATIVE

If we want to understand how children (and adults) develop their sense of public self, and the complexity of this human process, I believe that the anthropological theory of the person is a very useful place to start. In 1938, the anthropologist Marcel Mauss (1985) argued that there is a distinction between one's own sense of self (*moi*) and the social and cultural category of moral person (*personne*). The anthropological literature on the person, draws on Mauss' theory to show that even if different cultures have different understandings of the self and the person that are not rooted in Western individualism (Cohen 1994; Morris 1994), in all cultures, children and adults find themselves constantly negotiating their sense of self-distinctiveness to the group with the moral and cultural ideas of the person (e.g., the good child, the good woman, the good parent, the good Christian, the good citizen, the good activist, and so forth).

Influenced by Foucault's (2012) seminal work on subjectivity and Bourdieu's (1970) work on embodiment, the anthropology of the person also shows us that individuals learn these ideas through everyday social relationships and interactions with different social groups and institutions (e.g., family, school, friends), and that through self-construction individuals internalize (often unconsciously) the values and power relations of the society they live in (e.g., gender roles, class, ethnicity). I cannot dwell here in the complexity of these theories. Yet what I think we need to appreciate by referring to the anthropological literature of the person is that public self-construction is in fact a complex process of self-imagination, which unfolds in everyday life in the encounter with others (Escobar 2004, 252).

One of the most fascinating aspects of this process of self-imagination is represented by the fact that it often happens through storytelling. We tell narratives in public about who we are; and in these narratives, we negotiate with different forms of identification such as gender, culture, religion, ethnicity, political views, and many others (Pratt 2003; Escobar 2004; Alleyne 2000). At times, we privilege one form of identification over another. But most of the time, these different identities intersect in our individual stories, and we negotiate our belonging to different groups depending on life situations and depending on context.

Digital technologies have always played a major role in the construction of these narratives. The promise of internet technologies, and social media in particular, lies precisely in the fact that people can tell their stories; they can present themselves *in public* in a mediated and romanticized way and hence build their public selves. Yet this very promise is being challenged by our data-driven economies. Today we are no longer only digital citizens self-enacting and performing, we are all becoming datafied citizens whose data traces speak for and about us in public.

FROM THE DIGITAL TO THE DATAFIED CITIZEN

It is by considering the importance of storytelling and self-representation in people's lives that we can fully appreciate the social role of social media. In fact, social media have become a fundamental space for self-construction, a space where people can share their beliefs, build their networks, and tell public stories about themselves. This is why the sociologist Castells (2007, 2009) argued that with the rise of social media, we witnessed the rise of a new form of communication—the mass communication of the self—which equipped individuals with a new creative autonomy to express who they were and to bring about change in society.

It is for the same reason that the critical theorist Stiegler (2009) together with his *Ars Industrialis* collective argued that social media offered a new form of individuation, a possibility to "speak for oneself" and hence the key to empowerment, agency, and resistance. In the past, I argued that it is techno-utopian to believe that allowing individuals the space for self-expression necessarily brings positive change in society (Fenton and Barassi 2011). Still we cannot overlook that on social media, individuals tell stories about themselves in public, build their identities, and participate in society.

To describe the process of online identity construction and participation, some scholars have relied on the notion of *digital citizenship*. The concept of digital citizenship has been used in the past to describe how people participated in society through digital technologies (Mossberger et al. 2007; Dahlgren 2009) and that in doing so they "performed" their online self by enacting specific rights (Isin and Ruppert 2015; Couldry et al. 2014). Individuals thus become digital citizens online through daily speech acts (Isin and Ruppert 2015) and digital storytelling (Couldry et al. 2014) by claiming their own position in society and participating in communities of belonging.

Social media and other digital technologies have become key places of inclusion for these practices to flourish, and what makes them particularly appealing is that they are accessible also to those sections of society who are marginalized and excluded from voting rights, such as migrants or children. With reference to children, Third and Collin (2016), for example, have shown that children and youth often rely on these online spaces to confront, contest, and challenge the adult world in a public and performative way. According to Third and Collin (2016), children's ambiguous position in society, as not-yet-citizen, makes the performance of their digital citizenship incredibly creative and radical.

All these works on digital citizenship as a form of self-enactment are interesting because they shed light on the social role played by social media in the construction of the public self. Yet we need to ask, If one's own sense of public self is often constructed through narratives, then what happens when data traces "speak for" and "about us" in public? What happens when they tell a story about us that we do not recognize or identify with?

Under surveillance capitalism, we are not only digital citizens who self-construct online, we are datafied citizens because data traces constantly represent us in public, they tell stories for and about us that we often cannot control or rectify (Barassi 2016, 2017; Hintz et al. 2017, 2018). Children today are more exposed to this process because they are being datafied from before birth.

THE DATAFIED CITIZEN: WHEN DATA SPEAKS FOR AND ABOUT US

In winter 2019, I went for dinner with Cara a single mother of a 10-year-old who lives in Los Angeles. During dinner Cara told me that when she thought about digital profiling she felt as if someone was "talking" about her, and then she corrected herself:

C: No, actually I don't feel as if someone is only talking about me, I feel as if someone is gossiping about me. There is a fundamental differencc, you know. One day I was talking with the other mothers at school and a mother was telling a personal story about herself and her daughter, and then apologized for "gossiping." I told her that she wasn't gossiping.

V: Why not?

C Well because she was talking about herself. It's OK when you share information about yourself. That is not gossiping. Yet when others talk about you, when people seem to infer something about you on the basis of a certain information or rumor, then that is wrong, it feels like gossiping. When I get targeted for a search I have done on Google, it feels exactly like that, like someone has been gossiping about me.

As Cara was talking, I imagined data trackers as an army of gossipers that discuss specific choices we make and behavioral patterns behind our backs and use these pieces of information to come to conclusions about us and build stories about who we are. I then recalled the beautiful anthropological theory, which explores the everyday politics of gossip and how gossip can impact individual biographies in extreme ways (Besnier 2009). Whether we believe in associating data trackers to the very idea of gossip or not, I think that Cara's analogy is particularly important. It shows that one of the most problematic aspects of the ways in which data is being used under surveillance capitalism is represented by the fact that through profiling and predictive analytics, data technologies construct multiple, contradictory, and often entirely inaccurate narratives about who we are in public.

Data traces on their own are not problematic. The fact that one family buys a specific brand of nappies or shares the photo of their 5-year-old blowing out the candles of her birthday cake would hardly constitute an invasion of individual privacy or a direct harm for children. It is the *narratives* that are inferred from these data traces that make them problematic. These narratives are often used to define individuals in specific ways that could impact the construction of their public persona and the self-enactment of their citizenship. At the heart of this process of narrative construction is the very notion of digital profiling.

Digital profiling is often understood as a function of artificial intelligence that enables machines to bring different databases together and trace individual patterns (Elmer 2004). Yet, as mentioned in the previous chapter, it is also important that we map the multiple ways in which digital profiling is done within our datafied societies, not only by machines but also by humans, businesses, and organizations of all sorts who look up individual data traces to make assumptions about a person's life.

During the research, parents were starting to realize the implications of the narratives that could be built about their children on the basis of data traces. I often asked parents to imagine being a future employer or a principal of a new school and to describe how they would be able to profile their children on the basis of the information that the parent posted on social media. Many parents acknowledged that with the information that they posted on social media, people and machines would reach some key details about the family's political history, the family's social class, the health of children, interests, personality traits, gender identity, and so on.

When we think about digital profiling, we realize that one of the main transformations brought about by surveillance capitalism is that *data traces are made to speak for and about* us in public in ways that were not possible before. Educators in schools who believe in creating personalized education, bank or government officials who need to decide whether we can have access to credit or state benefits, prospective employers who need to choose us among hundreds of other candidates, all rely on digital profiling for decision making. As more and more narratives are constructed about us in public and made to act for us, we are all being turned into datafied citizens.

The datafication of citizens is of course not new. As mentioned in the Introduction, citizens have historically been profiled through their data (Hintz et al. 2018). Furthermore, according to the legal scholar Hildebrandt and Gutwirth (2008) the digital profiling of citizens can be dated back to almost twenty years ago and the early days of the internet. In fact, the proliferation of digital media enabled the construction of digital dossiers of a person's life, and these digital dossiers have long been used to determine whether an individual could have access to a mortgage or fly on a plane (Solove 2004).

Although digital profiling has a long history, the rapid extension of data technologies in different domains of everyday life has radically transformed its extent and implications for society. Today, data traces precede and speak for and about citizens in a variety of public contexts. Children are key to the understanding this transformation. They are the very first generation of citizens who are being coerced into digitally participating in society from before they are born, because their personal data is produced, digitized, shared, stored, analyzed, and exploited for them by others. The datafication of children thus does not only impact their right to privacy, but impacts their right to self-representation, moral autonomy, and contextual integrity (Nissenbaum 2010)

CHILDREN'S PRIVACY, MORAL AUTONOMY, AND CONTEXTUAL INTEGRITY

In her groundbreaking book *Privacy in Context,* Nissenbaum (2010) argued that when people are concerned about privacy, they are not really concerned about sharing personal data in public. What they want to have is a *control over their personal data flows* so they can determine which personal data is shared in which context. Most of the debates about privacy regulation with reference to new technologies are based on the assumption that there is a clear distinction between what is personal and should be private and what is public and should be visible (Nissenbaum 2010). Yet not only in our new data environments (and especially after the advent of social media) it is becoming increasingly impossible to divide the world between private and public, but too often our claims to privacy are dependent on context (Nissenbaum 2010).

There are multiple examples of this in everyday life. One might feel happy to share one's passion for wine and partying among like-minded colleagues yet feel uncomfortable if this information surfaced at a school meeting with other parents. One might feel comfortable sharing details about one's politics in one context and feel in danger in another context. It is for this reason that we need to understand that privacy is not only individual but it is social in the sense that our understanding of privacy is always shaped by the social environments that we live in, social norms, and conventions. Our claims to our right to privacy, according to Nissenbaum (2010), therefore need to be understood with reference to that very human need to control one's personal narrative in the different contexts (*contextual integrity*) and freely choose the values that we want to be associated with (*moral autonomy*).

Nissenbaum's contribution is of central importance because it shows us that a child's right to data privacy is directly interconnected to their right to self-representation. As we have discussed previously, the construction of one's own public self requires a complex process of self-recognition and self-distancing from specific relationships, especially the family. Individuals need to have the freedom to decide which inherited traits or qualities of their parents or families they want to endorse and which they do not. Maya for instance was brought up as a Christian and Republican, yet when she grew up, she distanced herself from those values and identified herself as a non-Christian Democrat.

Children should have the right to choose which family traits and values they do and do not want to identify with (right to moral autonomy) and to decide which type of information they want to disclose in specific contexts (right to contextual integrity). Yet both rights are being challenged by our new data environments. This is because today a plurality of agents are aggregating the data of children with the data of their parents and other family members and are profiling children on the basis of their family background.

Parents are aware of this. When I asked Amy, who lived in Los Angeles with her 5-year-old daughter and worked as a nanny, how a future employer or principal would be able to profile her daughter on the basis of the social media data that Amy produced, she answered

A: Well probably they would profile me first. They would identify me as liberal [laughs], so politically. It's funny because I don't post about political things, but they can get to that conclusion for the other stuff, the values that are important to me, the interests. It comes out.

V: How do you think your daughter would be profiled on the basis of the information you have out there?

A: They could see she is of Latino origin, and that she attends a Latino school (so I assume ethnicity) but then they would associate everything back to me (so also all the values, my income, and of course my politics).

Amy was right to highlight that the data inferred on social media about her social position, her values, and politics might come to define her child in public, simply through association. Of course children have historically been profiled through association with reference to their family's background, religion, politics, and culture. Yet in the past, there was less data available of one's childhood, and individuals had more freedom to dissociate from their family backgrounds. In addition to that, the process of profiling was not carried out by a plurality of algorithms and privately owned data technologies in obscure and unaccountable ways. As children are becoming the first generation of datafied citizens from before birth, parents are becoming anxious about what the future holds for them, and they feel that they cannot protect them.

CHILDREN'S DATA FUTURES AND PARENTS' ANXIETIES

In 2018 I interviewed Jill, who worked in marketing and lived in central London. She shared with me her dismay in view of the new data environments and her worry for children's future:

J: The situation at the moment is beyond creepy. They are manipulating people's behaviors. I think there are benefits to some of it [digitization], so for instance access to my health records. But I suppose my question is really: who else has access to those records? Are companies profiling patients? How are they using that data? So I see the benefits but I also think that regulation should keep pace

> with that. It's changed a lot in the last five years with the new regulations, but there are people that are always trying to get away with it and use data in ways that we can't really imagine.
>
> So the article on Facebook profiling about Brexit that I read, I was like, what? And the fact that somebody actually thought that up made me feel a bit sick. … You know the situation is out of control, if you think about babies, all the pregnancy and baby apps, they accumulate a lot of data and most probably they share it and profile children and market people.

Jill had a good awareness of the complexity of the data environments that we live in because she worked in marketing. I found it insightful that one of her first thoughts was directed toward children and the amount of data that was being collected from families since birth.

In this she was not alone. The more I talked to parents, the more I realized that when they thought about data privacy, they did not think about how data profiling impacted their daily rights (e.g., the prices they got, their premiums, credits, and so forth), they were more worried about the future of their children and how data traces would impact their children's lives. Of course they discussed their concerns with me because they were all aware about the nature of my research. But I think there are also different reasons to why parents (and society as a whole) seemed to worry more about children's data privacy than their own.

In an article written by the anthropologists Pink et al. (2018) it emerges clearly that in our new data environments, individuals feel that they cannot control their data flows, and for this reason are at times pervaded by data anxieties. Data anxieties, like other forms of anxiety, are often interconnected with the impossibility of knowing or controlling the future, and hence knowing or controlling how our personal data will be used (Pink et al. 2018). Children in a parent's mind often represent the future. In fact, young people and children are often negated in the here and now and they are referred to as "what they may become"; the future is therefore an imaginary construct that plays a fundamental role in the relationship between adults and children (Livingstone and Sefton-Green 2016). Hence, it is not surprising that parents focus their anxieties on children.

Many parents recognized that their children were being datafied in ways that were not imaginable before, and they were anxious about the possible ways in which this data could be used to define them and discriminate against them in the future. Their data anxieties often translated into different fears. Zack, the father of an 8-year-old and an 18-year-old, was worried that his children's data would be used by an over-controlling state in the future, or by corporations that could have access to genetic information. Nicole, the mother of an 8-year-old and 10-year-old in Los Angeles, who had a health condition, was worried that her children's data traces could be used by insurance companies to profile her daughters or by potential employers to discriminate them on the basis of their health. Although parents were clearly anxious about their children's data futures, they often felt that it was impossible to protect children's data privacy and that they were powerless.

THE IMPOSSIBILITY OF CHILDREN'S PRIVACY

One of the last parents I interviewed during my research was Charlotte, who lived in north London with her two children aged 6 and 8 years. Charlotte was going through a painful separation with her wife and was involved in a difficult custodial arrangement. She found herself being a single parent for most of the week and trying to juggle the requirements of a full-time job. As she sat down for the interview, one of the first things that she told me was that "she did not have the time or energy to think about data privacy." When I asked her whether she was worried about her children's data future, her answer struck a chord with me. It talked directly to the experience of many of the parents I met.

> C: I can't be bothered to be concerned. Maybe because I am naïve and I hope that everything is going to be all right. But I will probably regret it in 5 years or so. I am convinced that we will reach a tipping point and then people will feel outraged. For now I am sleepwalking and I could be doing a great disservice to my kids but in the short notice it makes my life much easier.

Like Charlotte, many parents I met during my research, understood the implications of children's data traces and worried that their digital choices were going to affect their children in the future. Yet they felt that in their immediate present they could not find the energy to be worried and did not want to change their digital use. Although Charlotte described herself as being naïve, Rachel for instance, who also lived in north London, was the mother of a 6-year-old and had also just gone through a separation. She told me that she "blocked the issue of data privacy out of her life." She was concerned about the future, especially in relation to the amount and type of data that was going to be available about her son. Yet "on a daily basis she deliberately chose not to think about it," because she felt that there was nothing that she could really do to change the situation.

If some parents, like Rachel and Charlotte, felt that they could not protect their children's privacy because they could not find the time or energy to worry about it, other parents that I met during my research, who actively tried to control their children's data flows, found that they also could not really protect their privacy. This is because, as I discovered in my own life, trying to protect children's data privacy in a datafied world where digital participation is often coerced (Barassi 2017, 2019) is not an easy task and involves daily battles, negotiations, and conflicts. Many of these battles we fail. I am reminded about this each day.

My life is filled with privacy desynchronies and failed attempts to protect my children's privacy. If one of my children is sick or falls, I search for health information on Google, even if I know that that information is going to be tracked. If I go to their pediatrician or if they have a health test, I do not read the terms and conditions about how their health data will be used, because I am too worried about their well-being. I never posted images of my children on Facebook, but I know that other parents have, and in few cases I did not ask them to take them down. I do not use voice-operated home hubs, but my children use Netflix, which gathers a lot of data about their viewing patterns and interests.

Whenever I think about the challenges I face in protecting the privacy of my children, I am reminded that I am not alone and I think about the parents I met during my research. I often remind myself about Annie and her husband who were being challenged by the other parents at their school because they did not want to sign the permission for photos to be taken at the school's Christmas pantomimes. I remember her anguish in standing her ground. I am also reminded about Linda's guilt when she told me that in the past she did not put her children's name on social media and she would ask everyone not to tag them in the photos. Yet she told me that eventually she slipped, and there are many photos online in which her children are identified by their first names.

Parents often feel that they can't keep up with the daily battles and efforts that are required to protect their privacy and the privacy of their children, and they feel powerless. Parental feelings of powerlessness can only be understood if we unpack one of the fundamental problems of current debates on data privacy: the focus on individual consent and responsibility. Current debates about data privacy are entirely focused on individual agency; when newspapers, researchers, and privacy organizations look for solutions, they focus on individual solutions. Journalists would offer advice on how to protect children's privacy; researchers find themselves developing privacy toolkits for parents and children.

In the age of surveillance capitalism, it is the individual (as a parent or anyone after the age of 13) who decides whether or not to agree to terms and conditions, and if they do not agree, tough luck, they cannot enjoy the benefits of the service. It is the individual's responsibility to set up privacy settings to make sure that companies do not track them, yet they often do nevertheless. It is the individual's responsibility to stop sharing photos on social media, to learn about encryption, and to delete browsing history. Yet at a historical time when companies and governments are largely benefiting from the ubiquitous, relentless, and pervasive tracking of our lives and when digital participation is often coerced, is privacy really only about individual responsibility and individual solutions?

I believe that if we really want to start addressing the social and political implications of our data environments, one of the first and fundamental steps that we need to take is to recognize that data privacy regulations are currently failing us and our children, precisely because they are focusing on individually centered solutions. We need to join forces and demand political change.

HOW DATA PRIVACY REGULATIONS FAIL CHILDREN

The General Data Protection Regulation that was enforced in Europe in 2018 is a great example of society's individualized attitude to data privacy and why individualized approaches to data protection are failing children. The GDPR was enforced on May 25, 2018, after two years of transitional period, and this implied that all companies (whether they were based in Europe or not) needed to abide to it if they were dealing with the personal data of EU citizens.

The main changes that the regulation has brought since the 1995 EU Data Protection Directive were the following:

1. A strengthening of individual personal data rights (e.g., all companies needed to make their practices more transparent, and users had to give explicit consent; they also had the right to be forgotten, and there were special provisions for the processing of sensitive data, biometric data, and children's data especially with reference to profiling); and
2. An increase in the fines for ensuring compliance (fines could go up to €20 million). Although the GDPR marks an important step toward data protection, we must understand that one of the main problems of data regulations like the GDPR today is the focus on transparency and individual consent.

The philosophy of the GDPR is pretty much based on the importance of transparency and choice. The basic understanding is that once data policies are more transparent and individuals agree to terms and conditions, companies then have the lawful right to process our information and the information of our children. Yet as Nissenbaum (2011) has shown, the focus on transparency of regulation leads inevitability to a contradiction of sorts, to a *transparency paradox*. In fact, simplifying privacy policies entails that many important and complex details, which describe the ways in which personal data is actually used, are left out.

Moreover, as argued in chapter 2, the focus on choice is deceiving because it does not take into account the multiplicity of times in which parents feel that they do not have a choice but to give their consent. What choice would they have if their most trusted doctor relied on an outsourced health portal with unjust terms and conditions? Or if their children's school suddenly decided to use Google Classroom?

As Savirimuthu (2015) rightly argued, the empowerment discourse about data protection that focuses on transparency and choice and assumes that citizens are agents in the protection of their own privacy (e.g., in requesting to be forgotten) does not address the social complexity of contemporary processes of datafication. It is for this reason that in May 2018, just after the GDPR was enforced, *Privacy International* criticized the new data regulations for focusing mostly on the personal data that is willingly disclosed by users without paying much attention that companies and governments are relying less on data we choose to provide and instead are looking at data they can observe, derive, and infer (Privacy International 2018).

Consequently, in our data-driven economies, parents do not have a choice or agency when it comes to data protection because their digital participation is often coerced and because their data is gathered by a plurality of agents and technologies beyond their consent and control. In the current data environments, therefore, the empowerment discourse is making parents feel powerless. Families often do not have the time or resources to file GDPR complaints or to approach all the companies that gather their data on a daily basis; they don't have the time or resources to enact their data privacy rights.

Another problem that families face is that even if they did have the time and resources to approach companies or file complaints, they would not necessarily be guaranteed that their requests for data privacy protection would be easily met. When the GDPR was introduced in Europe, for instance (Barassi 2018), I came across a European Commission fact sheet which discussed the "right to be forgotten," especially with reference to children's data. The fact sheet made it clear: "When children have made data about themselves accessible—often without fully understanding the consequences—they must not be stuck with the consequences of that choice for the rest of their lives," and as I was reading the fact sheet I nodded in admiration.

Then I read the next sentence, and my heart was filled with dread: "This does not mean that on each request of an individual, all his personal data are to be deleted at once and forever. If, for example, the retention of the data is necessary for the performance of a contract, or for compliance with a legal obligation, the data can be kept as long as necessary for that purpose" (European Commission 2017, 1–2).

I am not a legal expert and legal jargon in the majority of cases confuses me, but I was constantly asking myself, "Are they saying what I think they are saying?" Yet when I read the sentence I started to envisage all the different instances in my children's life that I think could fall under "performance of contract," and I came to the conclusion that children's "right to be forgotten" was not as straightforward as I thought it would be.

When we think about the failure of current data regulations, we also need to realize that data privacy laws are often difficult to implement. This latter point became clear when on June 20, 2019, the UK ICO (2019) published a damning report showing that most behavioral advertising in the UK is done illegally. Companies have been found to be profiling users by relying on sensitive information (such as health, religion, political views ethnicity, and so forth) without asking for explicit consent, which is a requirement of European law. The report also shows that companies can create extremely detailed unique ID profiles, which are defined by very granular and sensitive personal data and are repeatedly shared among hundreds of organizations without the individual's knowledge or consent.

Current data regulations thus do not protect us and our children from companies who are constructing unique ID profiles, which they share with a variety of organizations and are used to make data-driven decisions about our lives, without our consent or control. They also cannot protect us from the fact that these unique ID profiles are biased, inaccurate, and incomplete, because they rely on the inevitable *fallacy of algorithms* in human profiling. In our current data environments, as data traces are made to speak for and about citizens rather than focusing on the question about transparency and consent, regulators should start addressing and tackling the question about algorithmic fallacy and data justice.

CONCLUSION

This chapter has argued that one of the main transformations brought about by surveillance capitalism is that *data traces are made to speak for and about* us in public. Hence, we are witnessing the rise of a new type of public self, the datafied citizen. In contrast to the digital citizen, who uses technologies to self-construct in public, the datafied citizen is defined by the narratives produced through the processing of data traces; it is the product of practices of data inferences and digital profiling. The datafied citizen no longer owns his or her narrative in public and does not have any control over the narratives produced through digital profiling, even if these narratives are often used to make decisions about his or her life.

If we want to understand this transformation, children are the key. They are the very first generation of citizens who are coerced into digitally participating in society from before birth through the production, gathering, and processing of their data traces by others. Looking at the datafication of childhood can enable us ask what are the social and political implications of building a society where data traces are made to talk for and about citizens across a lifetime? How can we protect ourselves when we realize that the predictions and narratives that algorithms construct about individuals are reductionist, discriminatory and—in the case of children—preemptive representations of who they are?

8

DATA JUSTICE: THE HUMAN ERROR IN ALGORITHMS AND OUR DATA RIGHTS

I interviewed Lina on a summer day in 2017. The window in her living room was open and it overlooked the east London social housing estate in which she lived. It was a beautiful sunny day and we could hear the voices of the children playing at a distance. Lina migrated from Latin America to the UK more than ten years ago, and she shared the small apartment at the top floor of the housing estate with her two daughters, an 8-year-old and a teenager, and her husband. During the interview, I asked her how she felt when she thought about data tracking and digital profiling. I have never been able to forget her answer:

> L: When I think about all this surveillance I feel as if I were an object, like I was being constantly objectified. We do not have a choice, you don't have privacy, you don't have anything. I feel as if I am being belittled, minimized, and invaded. I feel little—how else can I explain it? I feel that it is too big for me, I can't fight it. I can't defend myself. I am completely powerless. I feel as if I am being used, because they could do whatever they want with your data and turn it against you.

As Lina was talking I could not but feel that she had perfectly summarized the injustice of surveillance capitalism. AI and data-driven systems objectify and belittle us. One of the big promises of big data and surveillance capitalism is found in the belief that human experience can be tracked, captured, and translated into data points and unique ID profiles, and that this data can be processed by "objective" algorithmic models that offer more precise and personalized solutions.

Yet, when it comes to human profiling, algorithms are always inevitably fallacious for three main reasons: they rely on inaccurate and decontextualized data (*algorithmic inaccuracy*), they are unexplainable and hence unaccountable (*algorithmic unexplainability*), and they are always biased (*algorithmic bias*).

Those who sell the promise of algorithmic accuracy or objectivity in human profiling are doing just that: selling a promise. What they are doing in actual fact is stereotyping people and presenting reductionist and simplified interpretations of their needs. Like Lina said, they belittle and objectify us.

Lina also mentioned that the data collected from her could be turned against her and that she could not defend herself. This is another fact of life under surveillance capitalism. In a plurality of contexts, from credit scores that determine whether one can rent a flat, to educational and health records that determine whether one has access to care or education, data can be used against us. There is always a human story behind these data records, and this story is often silenced by AI and other data-driven systems.

Even if all families are exposed to the data injustice and harm of surveillance capitalism, automated systems are usually more violent and unjust against people like Lina, because they often reproduce the systemic inequalities of society. Ethnic minorities, migrants, or low-income communities are often more exposed to data inequality because they find themselves trapped in a matrix of vulnerabilities (Madden et al. 2017) and in the mercy of automated inequality and bias (Eubanks 2018).

It is for this reason that we are witnessing the rise of new debates among the tech industry and policy makers that argue for the importance of combating algorithmic bias and developing fairer systems. These strategies are doomed to fail. When it comes to human profiling, AI and algorithms cannot be fixed, because algorithms are inevitably fallacious in capturing human experience and are always going to be biased. In the age of surveillance capitalism, we need to realize that the solution to data inequality cannot be a technological solution, but needs to be a political one. As more and more families and children are being datafied and exposed to the harm of algorithmic fallacy, we need to start to imagine what data justice (Dencik et al. 2019) would look like for families and children.

THE VIOLENCE OF DATA REDUCTIONISM

In a thought-provoking paper, Costanza-Chock (2018) considers how human identity and experience are violated and belittled by binary data systems and computer reductionism, which do not take into account the variety and complexity of human existence. Similarly to Costanza-Chock, I also believe that computer systems and algorithms cannot account for the complexity of human experience. The machines that we are building are actually machines that offer us simplified and reductionist understandings of human behavior; they stereotype individuals often by emphasizing personal weaknesses.

In everyday family life this data reductionism is experienced mostly through targeted advertising. Most of the parents I talked to defined targeted advertising as simplistic, inaccurate, or as missing the point. Dan for instance, who lived in London and had two children under the age of ten, told me that he was not concerned about data tracking because "algorithms don't really work." He also told me that he would start to feel scared if he had the feeling that they could profile him accurately and "really knew what he wanted."

Although most of the parents were not affected by targeted advertising, during the research I realized that some had very negative emotional experiences caused by the data reductionism of targeted ads. In her excellent paper on how people relate to the Facebook algorithm, Bucher (2017) highlighted the cruelty of algorithms and that the inferences that the Facebook algorithm makes about individuals "feel wrong" in everyday life, as the algorithm reinforces and reminds people of weaknesses that they would rather not like to think about.

During my research, different parents mentioned that being profiled felt wrong and affected them emotionally. Cara, who defined data tracking as a form of gossip also told me

C: Sometimes I am proud of the ads I get and sometimes I am like: why on earth are they targeting me? I don't deserve this. I am single, I never shared it, but I am targeted in that demographic.

V: Do you feel bothered by that?

C: When you ask if I am bothered, I feel that it is something that I should stand my ground. Being targeted as 50+ single lady bothers me. It embarrasses me.

Cara went on to tell me that being targeted and pigeonholed was upsetting especially because that targeting stereotyped her as a 50+ single lady. Those characteristics, she told me, were just aspects of her everyday life; they did not define her, and she preferred not to be reminded about them. Being reminded about them embarrassed her and hurt her feelings.

Cara was not the only parent who described that targeted advertising made her feel upset and embarrassed and affected her emotionally. I remember vividly a conversation with a friend, who gained more than 20 pounds as a result of her two pregnancies and was suffering greatly about her new body shape. She told me that she was constantly targeted for bigger sizes on Facebook and that—although she had tried to block the ads—the ads would pop back up after a while and upset her.

As I met parents, who shared their negative experiences of targeted advertising, I started to take note about the times in which I was targeted with ads that had a negative emotional impact on me. I remember one time in particular. A, my youngest daughter, was a few months old and exclusively breastfed. Breastfeeding—as many new parents find out—is not an idyllic and stress-free process; it is often a very complex process, which involves a great deal of self-doubt, insecurity, and personal crisis.

What made things particularly difficult for me was that in the afternoon I had less milk, and A spent hours screaming and failing to latch. At a certain point I considered the possibility of topping up with formula just for the afternoon, but I really did not want to, and I felt guilty. I looked for options online and I started to get targeted with formula ads. I wrote down in my fieldnotes, how I felt about those targeted ads:

> V: I think I might need to give A formula, but I don't want to and it makes me feel guilty. What really upsets me is that after I googled possible formula options, now I keep on being targeted with formula ads. This makes me feel so bad. It's deeply frustrating, tempting, and upsetting. To make things worse tonight on Facebook I was targeted by a different ad on baby probiotic that read: "Breast Milk. The perfect nutrition." I know breast milk is the perfect nutrition. But I can't do it, I really can't do it. I am trying ... but then there are all these formula ads. I don't want to see them. I feel awful. I don't understand how we can live in a society that monetizes on people's anxieties and thoughts; a society that boxes them and profiles them according to their own weaknesses.

Data reductionism and the constant targeting of people in family life on the basis of simplified understandings of their experience lead to all sorts of inaccuracies. Inaccurate profiling can come at an extraordinarily personal price. One of the most recent yet extremely revealing examples of the emotional consequence of inaccurate profiling comes from the experience of journalist Gillian Brockwell of the *Washington Post*. In December 2018 she described in an open letter to tech companies how painful it was for her to be targeted with baby ads, after her child was stillborn. In her heartbreaking letter, Brockwell writes:

> You surely saw my heartfelt thank-you post to all the girlfriends who came to my baby shower, and the sister-in-law who flew in from Arizona for said shower tagging me in her photos. You probably saw me googling "holiday dress maternity plaid" and "baby-safe crib paint." And I bet Amazon.com even told you my due date, Jan. 24, when I created that Prime registry. But didn't you also see me googling "braxton hicks vs. preterm labor" and "baby not moving?" Did you not see my three days of social media silence, uncommon for a high-frequency user like me? And then the announcement post with keywords like "heartbroken" and "problem" and "stillborn" and the 200 teardrop emoticons from my friends? Is that not something you could track? (Brockwell 2018)

Data reductionism and inaccurate profiling, as Brockwell's experience shows, can have a profound emotional impact. Once we realize that data reductionism in something that appears as mundane as targeted advertising can lead to emotional harm, then we need to take a step further and realize that in other areas of social life, from schools and health to government and policing, this reductionism is leading to profound real-life harm for individuals.

DATA HARMS AND INEQUALITY IN EVERYDAY LIFE

Crawford and Shultz (2014) in "Big Data and Due Process: Toward a Framework to Redress Predictive Privacy Harms," argued that we needed to consider the real-life harms of predictive analytics and appreciate that inaccurate and biased profiling could have a significant impact on an individual's life and livelihood. What the scholars rightly noticed was that as our societies become more and more datafied, we are constantly exposed to data harms, and the laws in place to protect us are still wildly underdeveloped and cannot keep pace with technological change.

To track data harm, Redden and Brand (2017), based at the Data Justice Lab at Cardiff University, launched the "Data Harm Record," a record of real-life harms that have been caused by uses of big data, automated systems, and AI. In their report, Redden and Brand (2017) highlight different examples of economic, social, and political harms created by data profiling.

Different cases mentioned in the report stuck in my mind for their unfairness. One example inspired by O'Neil's book, *Weapons of Math Destruction* (2016), referred to a high-achieving university student who kept being rejected from unskilled job positions and not being invited for an interview. When his father, who is a lawyer, decided to look into the issue, they realized that he was being discriminated for a mental health condition that the student disclosed in an employment personality test on an online application. I remembered the issue of educational data and how educational data brokers sell the data taken from online surveys. How many of these types of data harms are happening without us knowing? How many children are exposed to these harms? How many families?

Another example that Redden and Brand (2017) discuss in their report, and that shocked me for its irrationality and bias, was the individual whose credit limit had been reduced by American Express by 65 percent simply because he had been shopping in a retail store and a neighborhood where individuals had a low repayment history. The arbitrariness and bias of such a decision made feel angry, especially as I tried to imagine what that credit loss implied for his life. Of all the different examples that Redden and Brand (2017) mentioned in their record, what I found particularly upsetting was the reference to the data harm caused by institutional, police, and governmental profiling. In these cases, as Eubanks (2018) shows clearly in her book, data errors or bias can be devastating for family life.

Eubanks's (2018) book on automating inequality is particularly important because it shows that even if all families are exposed to potential data harm, there is something profoundly unequal and unjust about the different ways in which data harms impact white or high-income families on the one hand and low-income families or ethnic minorities on the other. Data technologies and automated systems are not equal or fair, and the experience of data harm depends on one's position in society.

This emerges well in the work of the legal scholar Gilman (2012) who shows that the poor are more exposed to privacy intrusions by government surveillance and other agents, and that current privacy law does not address the disparity of experience. Marginal communities are more exposed to privacy intrusion and data harm because in their everyday life they are subjected to systemic surveillance and discrimination. In addition to this, as Madden et al. (2017) have rightly argued and as the example of the American Express shows, poor and marginal communities are exposed to "networked privacy harms," because they are held liable for the actions of those in their networks and neighborhoods. It is because data and automated systems often amplify and reinforce existing inequalities in society that Eubanks

(2014) has argued that if we want to predict the future of surveillance, we need to ask poor communities.

During the Child | Data | Citizen project I had to come to terms with questions about data inequality and the disparity of experience in data harm, when I realized that understandings of data privacy in family life varied immensely. This difference in attitudes to privacy was not defined by the national/cultural context (US vs. UK) as I had originally thought when I started the research, but was always contingent to one's position in society and whether one had been exposed to inequality and injustice.

THE PRIVACY DIVIDE

Over and over again as I talked to parents I realized that those who were more concerned about data privacy and surveillance were those who were more exposed to social inequality, whereas those parents who "did not mind whether their data was being tracked" usually had a comfortable position in society. This is evident from the interviews with Mike, a white American who worked in local government, and Mariana, a Mexican immigrant who worked as a cleaner, both of whom I introduced in chapter 2. What struck me was how different their understanding of privacy actually was.

> Now it seems that everyone talks more about data tracking. It's definitely part of the culture. Now everyone knows that everything you do can be viewed by the NSA for example. But to be honest that really doesn't bother me, there is nothing that I am doing that I feel uncomfortable about, and if someone was to look at it or even broadcast what I do I would be OK with it. I mean it would be great if it could remain private, but if someone is viewing it and it became public, I mean, ... whatever. (Mike, white American, environmental assessor, Los Angeles)

> I don't agree with technologies; there is too much information out there. Privacy? Privacy does not exist. Now you have to be aware of the technologies, because you are also checked by the government when you pass the border; they check it and they can push you back. We are being checked by everyone, insurances, doctors, police, everyone knows what we do as a family, where we go, what we eat. (Mariana, Mexican immigrant, cleaner, Los Angeles)

In contrast to Mike, Mariana could imagine a plurality of agents that could have access to her family data, including government and border agents, and was worried about the impact this data would have on her life. Mike's and

Mariana's different perceptions of data privacy show that the understanding of the impacts of data flows and hence of the importance of data privacy is always contingent to the position we have in society. Mariana and Mike were not the only examples of privacy divide that I encountered during my research. When I asked John, a white British citizen with a high income, to describe his experience of daily digital surveillance and data privacy he replied that he "had no problem with it," and that he believed that dataveillance "makes people more honest." He then added

> J: You know people always lie to their insurers and other businesses and it makes me think that it is clever; this is the way it should work. I trust that data is generally factual, I think there will need to be regulations on how companies profile people, but I think that it is a good business strategy to profile individuals.

John's understanding of the impact of dataveillance was shaped by his comfortable position in society and that in thinking about profiling, he took the perspective of businesses over people. According to him, profiling and dataveillance were good for businesses because they made people more honest. What I found very interesting though was that he changed his understanding of profiling as soon as I mentioned web searches; then added, "No it is not OK to profile people for their web searches, you know I am OK with companies profiling me for my behavior, for what I do, but web searches are how I think." What I found particularly interesting in John's interview was that as soon as I mentioned one of his possible vulnerabilities, he shifted his understanding of data privacy.

Vulnerability was at the heart of people's different perspectives of data privacy. Those parents who felt more exposed to inequality or bias, and hence felt more vulnerable, had completely different understandings of profiling and dataveillance in comparison to those who, like John, felt that they were generally protected—until they felt exposed. A few weeks after I interviewed John, I interviewed Lina. As mentioned earlier in this chapter, Lina, in contrast to John, was very worried about data tracking and surveillance.

Those parents like Lina who actually experienced inequality on a daily basis, and thus understood that data could be turned against them, felt daunted by the massive extent of datafication and data harm. This emerged vividly in a conversation between Carlos and Jen, who had experienced discrimination and social inequality in their life:

J: I was telling Carlos the other day that we are going to be like in *The Matrix* that we are going to be controlled by all these algorithms, and you will think that you have choices, but you don't. You go on Google and you have all these advertisers, and so I was telling Carlos: "Why can't we be Neo, why can't we break out of the matrix?" You know the thing is that we see it. We know that they are collecting our information, but we can't stop ourselves from participating to this. Like you are so dependent on this. So, I kept asking my husband why we couldn't break out of the matrix and he answered: "Because we are not Neo."

Parents, like Jen, are quickly realizing that in the age of surveillance capitalism, data can harm and discriminate them. Yet they do not know how to tackle the problem of data inequality and injustice and how to protect themselves and their families. This feeling of powerlessness is not surprising because current data privacy debates and regulations are too individually centered and make parents like Jen feel like they need to act like Neo because it is their responsibility to protect their families.

Parents and individuals cannot combat data inequality alone; they cannot be held responsible for protecting their children and their family against the systemic inequality of AI and data-driven systems. What we need is a collective response. We need to join forces as businesses, organizations, and researchers and challenge institutional and business models that are shaped by the belief that algorithms are objective and can actually predict human behavior accurately. We need to demand a political solution and ask governments, businesses, and organizations to recognize that *algorithms cannot profile humans in just and fair ways*.

The research out there is doing precisely this; it is trying to uncover the bias of AI systems and their applications and to highlight how this bias is unresolvable. Dencik et al. (2016), for instance, have argued that a practice such as predictive policing faces three different (and unresolvable) challenges: the inclusion of preexisting biases and agendas in algorithms, the prominence of marketing-driven software, and the inability of interpreting and dealing with unpredictability. What the research is showing is that not only algorithms are profoundly racially biased (Noble 2018) but also that families are constantly profiled and subjected to automated systems that often reproduce social inequalities (Eubanks 2018). Digital profiling inevitably leads to biased, unjust, and misconstrued interpretations of people's lives, needs, and behaviors. Rather than believing in these promises of algorithmic efficiency when it comes to human profiling, we need to recognize that in the understanding of human behavior and psychology algorithms are *always and inevitably* fallacious for three main reasons:

1. They rely on imprecise and decontextualized data (*algorithmic inaccuracy*),
2. They are often unexplainable and unaccountable (*algorithmic unexplainability*), and
3. They are biased (*algorithmic bias*).

ALGORITHMIC INACCURACY: WHY ALGORITHMS CANNOT UNDERSTAND HUMANS

We are living at a historical time when our societies are being restructured by the cultural belief that data traces hold a key to human nature and can be used to profile individuals. But data traces are not the mirror of who we are. Humans think one thing but say another; feel one way but act differently. Mathematical deductions cannot understand the unpredictability of human behavior. We often do not use technologies as they are meant to be used. The unpredictability of our technology use leads to the production of broken, inaccurate, and contradictory data traces or to data traces that are deliberately tactical. During the project, for instance, I realized that many parents were becoming aware that they were being profiled by companies, and tactically tried to manage their profile either through practices of self-censorship or by "playing the algorithm." Zoe, one of the mothers I talked to, told me that she managed her profile by self-censoring. She wanted to make sure that all the purchases, actions, or beliefs "that did not look good" would not appear or be associated with her profile. For this reason, she was very careful about what she posted online, and she would choose which type of products to buy in shops and which things she would buy on Amazon. Most of her data traces were the product of this process of tactical curation of her profile and were not the expression of her beliefs or needs.

While Zoe tactically tried to manage her profile, Cara told me that she had "started to play the Facebook algorithm" to get a better experience and avoid being upset by the targeted ads that she received. Many times she would consciously choose not to like someone's post on Facebook, even if she liked it, because she realized that if she did not like things, her news feed became much more democratic and open to chance rather than likes. She also told me that she often tried to create a "happy day on Facebook" for herself. She had different tricks to do this. Either she would start liking the photos of animals posted by her friends, and she would get bombarded with cute animal feeds on Facebook. Or if she was having a bad day, she would do a web search for "2-bedroom house in Fiji" over and over again "just to be targeted with beautiful advertising of amazing places."

As the examples of Zoe and Cara clearly show, human digital practices are creative, unpredictable, and tactical, and this unpredictability of behaviors cannot be grasped by current data technologies. Hence, we are faced with a fundamental problem. Companies and data brokers use algorithms to process the data of individuals and sell the illusion that it is actually possible to translate human experience and nature in data points. Yet the data processed by algorithms is often the product of everyday human practices that are messy, incomplete, and contradictory; hence, algorithmic predictions are filled with inaccuracies, partial truths, and misrepresentations. Because we can trace connections and patterns does not necessarily mean that the knowledge we acquire from these connections and patterns is accurate, and boyd and Crawford (2012) explained this well when they suggested that:

> "Big data enables the practice of apophenia: seeing patterns where none actually exist, simply because enormous quantities of data can offer connections that radiate in all directions." (boyd and Crawford 2012, 668)

Of course, companies are trying to bring different forms of highly contextual data together under unique ID profiles (educational, health, home life data, social media data, and so forth); they are also buying data from a variety of organizations about our offline activities (e.g., our jobs, the car we buy, the mortgage we have, the insurance premium we share with family members). Despite companies gathering large quantities of personal data, the data they collect is always devoid from a thorough understanding of the human intentions, emotions, desires, relational contexts, fears, anxieties, and beliefs that informed it. In sum, the data that is analyzed is data that is systematically taken out of context (boyd and Crawford 2013, 669).

Algorithms are not equipped to understand human complexity and end up making reductionist and erroneous assumptions about the intention behind a specific web search, a purchase, or a social media post. Companies cannot tell if I search a specific product for me or for someone else, or if I like a post simply because I am tactically trying to change the algorithm, like I have done in many situations. Big Tech companies are aware of this and are trying to develop technologies that can actually access human thoughts. In March 2019, for example, in an interview with Harvard Law School Professor Jonathan Zittrain, Zuckerberg announced that Facebook was researching a new technology, a brain–computer interface that can access people's thoughts to navigate augmented reality (Harvard Law Today 2019). The problem for companies like Facebook that are trying to "tap into the human mind" is that as David Graeber once wrote, "human possibilities are in almost in every way greater than we ordinarily imagine" (Graeber 2007, 1) and cannot be predicted or anticipated by machines (at least not yet).

ALGORITHMIC UNEXPLAINABILITY: HOW CAN WE DEFEND OUR CHILDREN?

One of the biggest challenges faced today by parents and individuals is that in the majority of cases, the algorithmic predictions that are used to profile us and our children and to make data-driven decisions about our lives cannot be explained. According to O'Neil (2016), algorithmic models that are used in a variety of fields (e.g., insurance, policing, education, and advertising) are opaque, unregulated, and uncontestable, even when they are wrong.

This same understanding is also shared by the computer scientist Dan McQuillan in his piece about "algorithmic seeing." McQuillan (2016) believes that algorithms are the eye of big data; they are what gives meaning to the mass of information, but he also argues that algorithmic seeing is oracular rather than ocular (McQuillan 2016, 3). We are asked to have faith in algorithmic predictions like some people have faith in oracles, despite that these predictions cannot be explained and hence cannot be held accountable. Algorithmic predictions are thus obscure, opaque, and cannot be explained. Yet if algorithmic predictions are unexplainable and are defined by multiple obscure variables, then how can we guarantee that the type of predictions and decisions that they lead to are fair or accurate?

The issue of algorithmic (or AI) explainability is, at the time of writing, a fundamental issue of contention between privacy regulators and the industry. On the one hand, current regulations such as the GDPR are based on the belief that individuals have the right to explanation, especially when algorithmic decision making impacts their lives in negative ways.

On the other hand, within the industry there is a shared understanding that to focus on explainability challenges innovation. In an article in *TechCrunch*, titled "How (and How Not) to Fix AI," for instance, Joshua New (2018) from the Center for Data and Innovation, which defines itself as a think tank studying the intersection between data, technology, and public policy and has staff operating in Washington, DC, and Brussels, argues that regulators should not pursue algorithmic explainability, because

> The problem with this proposal is that there is often an inescapable trade-off between explainability and accuracy in AI systems. An algorithm's accuracy typically scales with its complexity, so the more complex an algorithm is, the more difficult it is to explain. (New 2018, para 5)

He thus concludes that seeking explainability would eventually damage AI businesses and stall innovation.

Although AI explainability may stall innovation, seeking explainability when it comes to human profiling is of central importance if we want to defend people's rights. If algorithmic predictions are unexplainable, how can we protect ourselves, our families, and children, and prove that they are wrong, when they actually are?

These questions are of course not new. Since the early days of the internet, legal scholars were concerned that citizens were being judged by automated systems that were nontransparent and nonaccountable, and decisions were being made about them that affected their lives without providing them with the right to due process (Solove 2004; Hildebrandt and Gutwirth 2008). Automated computer systems used for policies such as the "no fly" rule, adjudicated individuals in secret, without giving them access to a fair trial or allowing them the procedural protection of the law (Citron 2007). For this reason, Citron (2007) argued that we needed to introduce a new concept of technological due process, to enhance the transparency, accountability, and accuracy of rules embedded in automated decision making. In 2014, Citron and Pasquale (2014) argued that individuals should have the right to challenge decision based on scores that misrepresented them.

When we think about the industry-led claim that algorithmic and AI explainability stall innovation, we have to constantly remind ourselves that these technologies are used in a variety of areas of social life and that their predictions can cause real-life harm to people's livelihood. If we have to choose between AI innovation and defending our rights and the rights of our children, I personally would choose the latter.

ALGORITHMIC BIAS: REPRODUCING SOCIETY'S SYSTEMIC INEQUALITY

When we think about our data rights and those of our children, we need to recognize that algorithms are *always, at any-time* biased in one way or another. An algorithm is by definition a set of rules or steps that need to be followed to achieve a specific result. These rules or steps are never objective because they're designed by human beings and are the product of specific cultural values. This finding is of course not new. Friedman and Nissenbaum (1996) identified three types of bias in computer systems: *preexisting bias* (the bias of the humans that design computer systems and the bias produced by the cultural context that influences the design), *technical bias* (often there is a lack of resources in the development of computer systems, and engineers work with technical limitations), and *emergent bias* (society is always changing and thus the technologies designed at one given time or cultural context might become biased in a different time and context).

Although the understanding of bias in computer systems has a long history, the question about algorithmic bias has come to the fore especially in the last few years. In 2014, the Obama administration launched an enquiry into the impact of big data, which revealed that automated systems, although unintendedly, are biased and thus can reproduce existing forms of discrimination (Podesta 2014). By 2016, the issue of algorithmic bias exploded. American mathematician Cathy O'Neil (2016) in *Weapons of Math Destruction* argued that algorithmic models are biased and lead to data-driven decisions that reinforce racism and harm the poor. Barocas and Selbst (2016) called on the public, researchers, and policy makers to understand the disparate impact of big data on different sections of society. In 2018, two books were published that were crucial in framing these debates about algorithms, discrimination, and social justice: *Algorithms of Oppression: How Search Engines Reinforce Racism* (Noble 2018) and *Automating Inequality: How High-Tech Tools Profile, Police, and Punish the Poor* (Eubanks 2018).

What emerges clearly from these books and articles is that one of the great problems with data and automated systems is that, as Barocas and Selbst (2016) have rightly argued, these systems highlight specific patterns and present them as if they were useful regularities. Yet when we find patterns and regularities in social life, it does not mean that this data is in fact objective; on the contrary regularities in data may simply be the reflection of preexisting patterns of exclusion and inequality (Barocas and Selbst 2016).

Algorithms and AI systems are human made and will always be shaped by the cultural values and beliefs of the humans and societies that created them. It is for this reason that they amplify existing social inequalities and are inevitably biased. To say that they are inevitably biased implies that they cannot be fixed.

THE HUMAN ERROR: WHY AI SYSTEMS CANNOT BE FIXED

If we consider algorithmic inaccuracy, unexplainability, and bias, we quickly realize that when it comes to human profiling, algorithms *are always inevitably* fallacious. Consequently, if we want to tackle questions about the democratic challenges of AI innovation we need to critically reflect on the human error in algorithmic profiling and recognize that this error cannot be fixed. At the time of writing, more tech businesses and AI developers are trying to find solutions to fight algorithmic bias in their products and technologies. They are funding research and establishing advisory boards that are meant to scrutinize the ethical and political impacts of their technologies (e.g., AI ethics). At the heart of these strategies and practices adopted in the industry lies the very understanding that algorithms are biased because they have been fed "bad data"; hence, in order to rectify algorithmic bias, companies need to train algorithms with fair or unbiased data (Gangadharan and Niklas 2019).

Current strategies to combat algorithmic bias in the industry are profoundly problematic because they push forward the belief that algorithms can be corrected and be unbiased. Yet in doing so they amplify systemic bias. In fact, these strategies often rely on ideas of discrimination and fairness that are rooted in Western liberal thought and therefore exclude a wide variety of human experience. As Hoffmann (2019) has shown, for example, much of the current discourse on technological fairness is influenced by antidiscrimination discourse and the propensity of law to focus on one axis of discrimination at a time (i.e., gender or race). This does not take into consideration the intersectionality of the human experience of inequality (Crenshaw 1989), or in other words that gender, race, and other forms of identity are not separated in one's experience of discrimination.

Current strategies to combat algorithmic bias, in my opinion, are not only problematic but they are also completely missing the point about bias. There is no such thing as unbiased data. All processes of data collection require framing and processing (Gitelman 2013). Trying to combat algorithmic bias by believing that we need to train algorithms with good data that is fair and ethical, is a paradox of a sort, as it clearly shows that companies do not understand what bias actually is and how it operates. In February 2019, Yael Eisenstat, an ex-CIA agent who worked at Facebook for the business integrity division, argued that tech companies are tasking untrained engineers and data scientists with correcting bias, without having any critical understanding of how cognitive bias works (Eisenstat 2019).

Not only do companies lack an understanding of cognitive bias, but they fail to address the inescapability of cultural bias. Anthropologists have long been trying to grapple with how individuals necessarily interpret real-life phenomena according to their cultural beliefs and embodied experience (Clifford and Marcus 2010), and that cultural bias necessarily translates into the systems that we build, including scientific systems (Latour and Woolgar 1986). From an anthropological perspective, there is nothing that we can really do to correct or combat our cultural bias, because it will always be there. The only thing we can do is to acknowledge the existence of bias through self-reflexive practice and admit that the systems, representations, and artifacts that we build will never really be objective. This same understanding should be applied to algorithms.

Current debates on algorithmic bias, AI, and ethics thus may be doing more harm than good in addressing the problem of data inequality. This is because rather than highlighting the issue of the *inevitable* fallacy of algorithms when it comes to profiling humans, these debates seem to be placing algorithmic decision making and automated systems in a more positive light by emphasizing the industry's commitment to fairness, accountability, and transparency (Overdorf et al. 2018; Gangadharan and Niklas 2019). Human error is an inescapable dimension of algorithms and is something that cannot be corrected. If we are really serious about tackling bias, then we must realize that the solution to data inequality cannot be a technological one but needs to be a political one.

THE HUMAN ERROR AND THE STRUGGLE FOR DATA JUSTICE

Human error will always define algorithmic profiling. Yet today companies, data brokers, and governmental institutions collect, sort, analyze, and process different typologies of personal data to map them to unique ID profiles. They then use this data to profile citizens, build narratives of people, and make data-driven decisions about them. These profiles are flawed because they are shaped by the inevitability of *algorithmic fallacy* when it comes to human profiling, yet they are made to speak for and about individuals.

In thinking about how data is made to speak for and about individuals in data-driven and automated decision making, the legal scholar Solove (2004) believed that we needed to move away from a focus on surveillance and the metaphor of Orwell's Big Brother (Orwell [1949] 1983), and we instead needed to refer to the Kafkian metaphor of *The Trial* (Kafka [1925] 2020). I remember reading Kafka's *The Trial* in high school, and I remember the disbelief and anguish as I followed the story of Joseph K, the protagonist, who was arrested and prosecuted by a remote, unknown, and inaccessible authority without him (or the reader) knowing what he had done wrong.

In Kafka's *The Trial*, the author meant to depict "An indifferent bureaucracy, where individuals are pawns, not knowing what is happening, having no say or ability to exercise meaningful control over the process" (Solove 2004, 37). Solove believed that digital profiling was subjecting individuals to the same process; individuals' personal information was judged and processed by "Databases with little intelligent control or limitation, and individuals had no way to contest the decisions made about them" (Solove 2004, 39).

His work seems to be extraordinarily premonitory of today's data environments. In fact, today, while it is clear that we are being constructed as data subjects and these constructions are impacting our lives in a variety of ways, it is also clear that the process of algorithmic definition and construction escapes our control (Cheney-Lippold 2017). As data traces are used by data brokers and other agents to build narratives about who we are, we often do not have access to these narratives, we can't rectify them, challenge them, and tell our version of the story, even when these narratives are wrong.

In our current data-driven economies, the data subject is much more than an individual consumer or user sorted by one's own choices whose privacy has been compromised, it is a citizen whose freedoms and rights are highly dependent on the ways in which he or she is being profiled. As Dencik et al. (2016) argued, we can no longer only focus on data privacy, but we need to start talking about data justice and unpack how surveillance and datafication are tightly interconnected with questions about social justice (Dencik et al. 2016). This implies that we can no longer focus only on how an individual's data is collected, stored, or algorithmically processed, but we should rather explore how data-driven decision making is part of a particular economic and political agenda that seeks to systematically stigmatize, marginalize, and exclude minorities (Metcalfe and Dencik 2019).

In the last few years, the concept of data justice has been used to expose the systemic inequality of automated systems and big data (Taylor 2017) and to highlight the power relationships involved in the development of data-driven decision making (Heeks and Renken 2018). At the time of writing more work is being done in the field, especially by research hubs like the Data Justice Lab at Cardiff University and the Data & Society Research Institute in New York City. Clearly we can no longer focus on the ways in which individual data is aggregated and processed, but we need to question the ways in which data systems and technologies are built and operate and how they are actually amplifying existing social inequalities. In other words, we need to start acting collectively and imagine what data justice would look like for us and our children.

DATA JUSTICE FOR CHILDREN?

In June 2019, my husband, Paul, and I were chatting to a couple of parents at P's preschool culmination day in Los Angeles. They asked us about our recent trip to Disneyland, which was part of P's fifth birthday celebration. Paul said that we had a great time, but that we were outraged when we discovered that Disneyland has introduced a new facial recognition ticketing system. He described how frustrated and powerless we felt when P's face was scanned at the gate without giving us any notice or privacy policy to agree to and after we bought the tickets. Paul was passionately talking about the issue; over the years he has totally embraced my cause. He went on to discuss our anger when the attendants would not give us a clear answer or tell us where we could find a data policy detailing how children's face scans were going to be used. As Paul described our frustration, Sylvia, the mother of one of P's classmates, looked increasingly confused:

S: Why do you mind so much? What is the problem?

V: Well would you be OK if they fingerprinted your children at the entrance to Disneyland? You know, biometric data can be extracted from a face scan, so it's the same thing as a fingerprint.

S: I wouldn't have a problem. Most probably the data is only used for advertising purposes.

V: Most probably. But there is no way for us to know. Right?

I went on to explain about my project and my book and described the datafication of childhood and my findings. I tried to build my case by explaining why it was wrong for companies to gather so much data of children, especially biometric data. I told her that I believed that something needed to be done, that we needed to raise awareness and make sure that greater protections were in place. Sylvia, listened carefully and then she surprised me when she said

S: I am ignorant, I admit it. I have no idea of what is going on with children's data. But maybe you are ignorant too. Maybe you miss the full picture and do not realize that there is nothing that you can actually do. It is happening everywhere; there is no way to stop this.

She had a point. I do not have the full picture of what happens to children's data, and the situation is changing rapidly as more and more businesses and institutions are relying on the systematic collection, archiving, and processing of children's personal information. Although I do not have the full picture, I strongly believe that something can be done, especially if we start imagining what data justice looks like for children and families.

I imagine the struggle for data justice as one that focuses on three necessary changes: a policy change (a change in data regulations), a political change (a change in institutional models and organizational practices), and a cultural change (a change in current perspectives of data privacy).

To change policy, we need to design regulations that take into account the complexity of our current data environments, regulations that are actually able to address three main problems when it comes to children's data privacy:

1. The coercion of digital participation: We need regulations that move beyond the discourse of individual choice and individual responsibility and actually hold businesses and organizations accountable for the technologies and data collection practices that they use. Regulations like the GDPR or the California Consumer Privacy Act of 2018 stress that businesses and organizations need to make their data collection practices transparent and provide consumers with the possibility to opt out or be forgotten. Yet

these regulations are very difficult to implement, and the main focus of such policies is on families' right to know and right to complaint. Families often do not have a choice, and in the majority of cases they do not have the time or resources to file complaints. These regulations are thus doing a disservice to families and children. A radical change in this area is needed. Regulations should abandon the focus on transparency and individual responsibility. What they should do is enforce privacy by design and make sure that opt-out models are replaced with opt-in models; for example, I should have been offered the facial recognition ticketing service at Disneyland, not had it imposed on me.

2. Aggregated profiles: Children's personal data is collected, stored, and processed through adult profiles or aggregated household profiles (e.g., on social media, or home technologies), which do not have to abide by children's data privacy regulations. All the children's data that are collected through these technologies should be deleted and not processed. New regulations should make sure that children are not judged or profiled on the basis of the families and collectives they are brought up with, and household profiling should not be allowed.
3. The role of data brokers and the creation of unique ID profiles: The world of data brokers, data sharing agreements, and digital profiling is extraordinarily complex and impossible to grasp. We need to start investing in public and independent research that tackles their practices and provides useful tips on how we can regulate the sector. We also need to start challenging business models that do not understand the fallacy of algorithms when it comes to human profiling and those that stereotype and discriminate individuals on the basis of a false promise.

Fighting for children's right to data justice implies not only that we struggle for a change in data regulations and policy, but that we join forces together to push for a political change in the ways in which automated systems and AI are increasingly being used to profile and govern citizens. Citizens are governed through data in ways that were not possible before. Predictive analytics is used by policing and courts; biometric monitoring is a common practice by border patrols; and data-driven decision making is used by governments to decide fundamental matters such as welfare provision or child protection. As a society we need to challenge these practices.

We need politicians who are ready to understand the *human error in algorithms* and that data-driven services reproduce systemic injustices and social inequality. If we want to protect children and citizens, we need to find political solutions that recognize the profound social implications that emerge with the datafication of citizens.

Last but not least, I believe that fighting for children's right to data justice also implies that we realize that much of our understandings of privacy and big data lies on Western, white, middle-class demographics (Chakravartty et al. 2018; Arora 2016, 2019). Hence if we want to ask questions about data justice we need to take into account the global south (Milan and Treré 2019; Arora 2016) and to consider how the world is being structured into a new type of social order, the one of data colonialism, where data is taking over and capitalizing everyday life across different cultures (Couldry and Mejias 2019).

One of the main limitations of the Child | Data | Citizen project is its Western-centric perspective. I am sad that due to a lack of time and resources I have not been able to tackle critical questions about how different countries are datafying their children. Yet if we really want to start fighting for children's right to data justice, we need to start from there; from an in-depth, cross-cultural analysis of data governance worldwide. If we do so, we may be able to grasp the complexity of the datafication of children and to find political solutions that limit the harm of surveillance capitalism.

CONCLUSION

Individuals are being datafied from before birth. They are being tracked and profiled by algorithms that look for patterns and regularities among messy, inaccurate, and decontextualized data traces. These algorithms that are used to capture and understand human experience and behavior are always inevitably fallacious. They are fallacious because they are biased, inaccurate, and unexplainable. The human error in algorithms inevitably leads to data harm, because automated systems profile individuals in nontransparent and unaccountable ways.

Even if all families are exposed to the data harm created by the human error in algorithms, we must acknowledge that the poor, ethnic minorities, and marginal communities are often more exposed to data violence, systemic surveillance, and bias. Once we understand the deep-seated social inequality of data-driven decision making, we can appreciate that we need to join forces as researchers, businesses, and policy makers, and we need to start imagining institutional and business models that abandon techno-utopian dreams of automated efficiency and address the fallacy of algorithms. In other words, we need to start fighting for the right to data justice.

9

CONCLUSION: AI VISIONS AND DATA FUTURES

On Christmas 2018, our neighbor and friend bought A, who at the time was one and a half years old, a dancing robot toy designed by Fisher-Price. I was delighted. In the age of internet-connected and artificial intelligence toys, the smiley robot was not only very cute but was also very basic in terms of technological affordance. It danced and sang; it taught her colors and made her play. It also offered the option to press a red button and record sentences. The robot would play them back with the music. That is about it. A unwrapped her gift, looked at it for few minutes, and then lost interest. P, who at the time was 4 years old, loved it.

P is bilingual and so she started to audio record sentences in both English and Italian and waited for the robot to play them back. Then she tried to see if the robot could understand Spanish as well, and so she recorded few sentences that she had learned at school. Then she looked up at me with excitement, and said, "Mamma he is so intelligent." I just nodded and smiled. Yet I should have said, "No, you are intelligent," but I did not want to ruin her excitement. The interaction with P brought me back to when I was child. I always dreamed (and still do) to have robot that I can interact with—a robot that becomes some sort of life companion, that amuses me with its intelligence and cuteness, and that takes care of things for me. When P looked up at me with her excited grin, I reminded myself about my own dream and did not want to crush her magical moment.

The interaction with my daughter, however, reminded me about Meredith Broussard's book, *Artificial Unintelligence* (2018), which argues that much of current understanding of artificial intelligence is rooted in an unfounded perception of "intelligent machines." In the book, she mentions the case of AlphaGo, the AI program that is often cited as the perfect example of intelligent technology, because in 2017 it defeated the world chess champion, not one but three times.

Broussard (2018) argues that AlphaGo is a remarkable mathematical achievement; it contains data from 30 million games and processes that data to play the game more effectively, and win. Yet she also argues that the technology is not intelligent, it does not have a consciousness or autonomous thinking, nor does it fully understand the symbols that it manipulates. The intelligence of AlphaGo, she argues, is the intelligence of those who designed the technology.

There is much similarity between my daughter's belief in the intelligence of the little dancing robot and current understandings of AI. This is because current visions of AI solutions and innovation are a perfect reflection of the techno-fetishist and Western-based cultural imagination of intelligent robots, which has a long history. I will explore current AI visions, and I will tap into those shared popular beliefs that see AI technologies as capable of solving society's most problematic issues, as well as those collective fears and anxieties that see these technologies as being here to destroy humanity and take over the world. AI technologies are not here to save or destroy us but may be radically transforming the world as we know it. At the heart of this transformation, I will argue, we find once again the very question about our data futures, algorithmic fallacy, and data justice.

AI VISIONS: FROM MYTHS TO REALITIES

In June 2015 an article appeared on the science and technology website Gizmodo titled, "Artificial Neural Networks Can Day Dream—Here is what they See" (Campbell-Dollaghan 2015). The article discussed Google's announcement that they were experimenting going deeper in the neural networks. The experiment involved asking their AIs to enhance what they saw in a specific image by detecting not only shapes and edges but also elements, and then to build on that information. In one case they fed an image of clouds through a layer that had already been trained to detect animals, and the result was astonishing. Like it happens when we look for animal shapes in clouds, the AI produced the image of a sky populated by animals.

The company went even deeper with the experiment. It placed the neural network in an endless feedback loop with an image it generated. The results are visually incredible. One network—expert in arches—reproduced the arches thousands of times in the image and, in doing so, it built an image of a magical world of colors, lines, and doors. Google defined this practice *inceptionism*, and the images produced *dreams*.

I was amazed and terrified. The thought that AIs could dream made them feel so real, so humanlike and fascinating. At the same time, it made me feel anxious and worried about the type of tech-beings that we were creating and the implications for society. Of course, I quickly curbed my excitement and anxiety and told myself that AIs cannot dream. As far as I know, dreams are the product of the unconscious; for AIs to be able to dream, they would first of all have to be conscious.

Although I curbed my own excitement and anxiety, I concluded that the article, and my own reaction to it, have a lot to say about our very Western fascination and relationship with artificial intelligence. In fact, the article is a great example of some of the current cultural narratives defining AI that construct these machines not only as if they were humanlike (e.g., able to dream) but as if they could go beyond humans and enable us to grasp some of the mysteries of the human mind (e.g., their dreams can be captured and visualized). If the article was a great example of the current cultural imaginations of AI, my own reaction reflected the fears and anxieties that the image of intelligent machines brings to mind.

The tension between dream and fear, excitement and anxiety has defined the entire history of artificial intelligence, which is polarized between positive and negative perceptions (McCorduck 2004). This is really unsurprising. Especially in Western cultures, *all* understandings of new technologies have traditionally been polarized between techno-utopian and techno-dystopian visions. In fact, as Segal has argued (1985), Western social thought has not only been dominated by utopian understandings of technological progress, which see technologies as directly interconnected to new forms of social emancipation and justice, but also by techno-dystopian understandings of technologies as tools of suppression and control.

From an anthropological perspective, what is fascinating about these techno-utopian and techno-dystopian visions is that they involve a process of *technological fetishism* (Harvey 2003). Humans tend to invest technological objects with a specific form of power, almost magical power, and they believe that these objects are able to move and shape the world (Harvey 2003, 3). Although fetishism is a human process that can be found in a variety of cultures (Hornborg 1992; Graeber 2007) technological fetishism is often at the very heart of Westernized notions of modernity and progress (Hornborg 1992; Pfaffenberger 1988). The Western fascination with the newness of technologies has enabled us to construct mythical understandings of technologies (Mosco 2004) and to perceive these technologies as if they were magical, capable of transforming our societies (Morley 2006).

We cannot understand the rise of artificial intelligence without referring to these technological myths (Natale and Ballatore 2017). Yet when we think about the current cultural narratives and myths that shape the discourse around AI, we need to realize that these discourses are not simply emphasizing the advent of new technologies—like the telegraph or satellite and internet technologies—that can transform the organization of society. What they are clearly emphasizing is the advent of a "new phase of human existence," one in which humans and machines would not only coexist, but also dialogue, empathize, or even in the worst-case scenario, run into conflict (Bory 2019).

Ideas of newness, transformation, innovation, and radical change dominate contemporary understandings of AI solutions, and these understandings at the moment are shaped by great expectations and fears, precisely because these technologies are seen as enhancing or destroying humanity. But once we appreciate that these hopes and fears are rooted in techno-utopian and techno-dystopian imaginings of new technologies, we need to ask ourselves how realistic AI visions actually are? Are we really building intelligent machines? How far have our dreams and nightmares of artificial intelligence come?

THE INTELLIGENCE OF AI?

In June 2017, the story broke in the news that researchers at Facebook had to shut down an experiment after two artificial intelligent robots appeared to be chatting to one another in a strange language that was not understandable to humans. The news was inspired by a blog post written by researchers at Facebook that discussed an experiment to teach AI systems to negotiate between one another, as if they were trading like human beings. Different tabloid newspapers, blogs, and websites reported that Facebook engineers panicked when they realized that the AIs had created their own language that was not accessible to humans. They defined this event as scary, creepy, and as a turning moment in AI innovation, a moment when AIs had started to talk to one another.

The news coverage was, as it often happens, sensationalist, inaccurate, and absurd. In the first place, there was nothing really new in AI-to-AI communication. Google and OpenAI have run different experiments in this area. In the second place, it is not true that Facebook engineers were spooked by the AI technologies communicating in a strange language and or that the experiment was shut down. The programmers realized they had made an error by not incentivizing the chatbots to communicate according to human-comprehensible rules of the English language. In their attempts to learn from each other, the bots thus began chatting back and forth in a derived shorthand (McKay 2017).

Although misconstrued and sensationalist, the story of the AIs talking to one another is the perfect example of the ways in which, in popular discourse and the press, there is a tendency to overestimate the intelligence and autonomy of AIs and their potential dangers. Yet in the age of machine learning, algorithmic logics, and big data, the question of whether we can actually build intelligent machines is still largely unanswered.

The question about intelligent machines has a long history and finds its roots in a cultural enquiry that dominated Western thought for centuries. The Western philosophical tradition has been defined by the will to grasp the difference between nature and culture, mind and body, and whether we can replicate the natural systems (e.g., the brain) in mechanical artifacts. This history of thought is beautifully explored by George Zarkadakis (2015) in *In Our Own Image: Will Artificial Intelligence Save or Destroy Us?* He explores these questions from the ancient Greeks to Descartes and Hobbes and shows how within Western philosophy there has always been a tendency to reflect on the human as a complex machine, one that could possibly be replicated.

Zarkadakis also shows that throughout history and across different civilizations, scientists have consistently tried to build automata, mechanical beings able to assist or defy humans. One famous example was the Mechanical Turk, a chess playing machine that was built by Wolfgang von Kempelen in 1770 to impress the Empress Maria Theresa of Austria. The Mechanical Turk was in fact a mechanical illusion that allowed a chess player to hide in the machine. Yet it was a remarkable example of the search for the mechanical mind.

Even if the history of Western thought has been dominated by the will to replicate natural processes in mechanical artifacts and build automata that would assist and supersede humans, it was during the 1950s that the very question about intelligent machines came to the fore. In the 1950s, Alan Turing the British mathematician who is heralded as one of the fathers of artificial intelligence, argued that it was possible to prove the intelligence of machines with a simple experiment, called the Imitation Game. Turing's Imitation Game took place in an imaginary house with three rooms, each connected to one another via computers. In one room we have a man A, in one a woman B, and in one a judge C. Basically the game consists in the fact that A needs to convince the judge that he is the man and that B needs to deceive the judge that she *is* the man (Zarkadakis 2015). According to Turing, if someone replaced B with a machine and the machine was able to deceive the judge, then the machine was intelligent:

> The machine would imitate the man: when asked whether it shaved every morning it would answer "yes" and so on. If the judge was less than 50 percent accurate in telling the difference between the two hidden interlocutors the machine was a passable simulation of a human being and, therefore, intelligent. (Zarkadakis 2015, 48)

In the 1980s John Searle slammed Turing's Imitation Game by suggesting that because a machine is instructed to deceive does not imply that the machine is actually intelligent. He created his own experiment, known as the Chinese Room. The basis for the experiment was exactly similar to Turing. However, Searle introduced a new dimension. In fact, in his Chinese Room, the messages were exchanged in Chinese. Searle noticed that when a system received an input in Chinese, it was able to match that input into an output that was also Chinese. Yet the machine did not necessarily know or understand Chinese. His experiment led him to the conclusion that the machine could process an input, follow logical instructions, and deceive the judge. Yet there was no understanding and consciousness of the symbols manipulated. So, according to Searle, we could not really talk about intelligent machines because, "Without consciousness there is no true intelligence" (Zarkadakis 2015, 52).

Searle's critique to Turing has been often contested (Cole 2019). Yet when I read about the Chinese Room and that the manipulation of symbols cannot really be understood as an understanding of symbols, I was suddenly thrown back to my readings in anthropology and symbolism. The way in which humans understand, internalize, and act on the basis of symbols has been a key area of investigation in anthropology. In anthropology, it is clear that there is a great complexity to the ways in which symbols work for human intelligence and experience. One example can be found in the work of the anthropologist Victor Turner, titled *The Forest of Symbols* (1970), which is particularly insightful.

By focusing on the multilayered and condensed aspects of ritual symbols, Turner argued that symbols often have two different meanings for humans:

one that reflects a moral and normative aim ("the ideological pole"), and one that reflects people's sensory experience, emotions, and feelings ("the sensory pole") (Turner 1970, 30). When I try to explain this to students, I often use the example of the Italian flag. I am Italian, and when I look at the flag I am reminded of the nationalism and patriotism of Italian culture with which I often do not identify, but I am also reminded of Italian cultural values that I respect (normative and moral poles). At the same time, the Italian flag speaks directly to my sensory experience by reminding me of my mom, my friends, and all the things I loved from childhood. My understanding of the Italian flag as a symbol thus is rational and emotional at the same time; it speaks directly to my lived, embodied experience as a human.

Even if Searle's experiment is still open to question (Cole 2019), what he clearly asked is whether machines can understand symbols like humans. Keeping in mind the anthropological knowledge about the human complexity of symbols, what would it take to train a machine to gain that embodied, emotional, and human understanding of the symbol? I concluded that there is no way that current AI technology would be so sophisticated, developed, and humanlike enough to be able to grasp that anthropological nuance.

As much as we are surrounded with news stories like the one of Facebook bots being too intelligent, presently AI cannot replicate human intelligence. In this understanding I am not alone. According to Broussard (2018), popular and business discourse is dominated by a confusion of sorts when it comes to understanding the difference between *general AI* and *narrow AI*. General AI is the Hollywood version of AI, that idea of a humanlike AI able to feel emotions and deceive, an AI with consciousness and autonomy. Narrow AI instead works by analyzing an existing data set; it learns from these data sets and makes predictions. Machine learning, neural networks, and predictive analytics are all examples of narrow AI. Broussard (2018) believes that in the business and popular discourse there is much confusion between the two different types of AI, yet general AI does not exist, whereas narrow AI is what we have.

Current AI is not sophisticated enough to come anywhere near our cultural imaginations of artificial intelligence. The problem is that not only we do not have a full grasp of the functioning of the human brain and hence cannot faithfully reproduce it, but we also do not have a full grasp of the complexity of human experience, of what we understand as human consciousness, and what makes us the way we are. Until we have a full understanding of both, it is impossible for us to create the artificial intelligence of our cultural imagination. In addition to that, even if eventually we came to a deep understanding of both—the human brain and human experience—this does not imply that we would be able to replicate them into a mechanical abstraction.

The understanding that we have not yet created the machines that are the product of our cultural imagination, machines that can replicate and dominate us with their intelligence, is key. It can enable us to deconstruct current techno-fetishist models of AI and to see these technologies and business discourses for what they really are: a human product. By doing so, we can actually focus on the ways in which AI, in the narrow and current form, is transforming society. Only by focusing on what we actually have and the changes that are happening to our society as a whole can we start anticipating future risks.

AI AND DATA FUTURES

The risks of AI can be great. One year before his death, Stephen Hawking argued during a talk at the 2017 Web Summit Technology Conference in Lisbon that the potential of AI in combating climate change, disease, and poverty could be immense:

> Unless we learn how to prepare for, and avoid, the potential risks, AI could be the worst event in the history of our civilization. It brings dangers, like powerful autonomous weapons, or new ways for the few to oppress the many. It could bring great disruption to our economy. (Hawking in Kharpal 2017)

One of the dangers in AI innovation lies in the question about human profiling and algorithmic fallacy. We are living at a historical time when every little detail of our lived experience is turned into a data point from before we are born. Private companies are collecting multiple typologies of an individual's data (home life, social media, health, and educational) and are aggregating these into unique ID profiles. This information is processed by AI technologies to profile, judge, and make decisions about us. Governments, businesses, and institutions of all kinds are relying on AI to make key decisions about individual lives in a plurality of areas from recruitment to policing. Hence, we cannot really address the questions about AI futures without critically engaging with the questions about algorithmic fallacy and data justice.

As we have seen, the real problem of our times is that when it comes to human profiling, algorithmic models are inevitably fallacious because of their intrinsic human error (bias, inaccuracy, unexplainability). As a society however, we are buying into techno-fetish dreams of AI innovation, and we are failing to recognize the fallacy of algorithms when it comes to human profiling. Wherever we turn, AI technologies are being constructed by businesses and governing institutions as the perfect solution to profile humans and to solve a plurality of problems, from finding the right person for our workplace to preventing crime. They are trusted with fundamental decisions that impact individual lives.

Yet, algorithmic predictions when it comes to human profiling are always inaccurate, unexplainable, and biased. Algorithms cannot really understand humans in a just and fair way; they are designed by human beings and are the product of specific cultural values. On top of that, the data traces collected and the databases created are always inaccurate and embedded with cultural meanings and understandings. So when machines learn, they learn not only from biased algorithms but also from biased data. Examples are many and some of them are terrifying. For instance, in May 2019, researchers at the AI Now Institute in New York published a terrifying report on the uses of AI technologies in policing. The report revealed that in different jurisdictions across the US, the AI technologies used by police for predictive policing relied on *dirty data*, police data that was produced through historical periods of flawed, racially biased, and at times illegal practices and policies (Richardson et al. 2019). In the report, the researchers demonstrated that if predictive policing systems are informed by such data, they cannot escape the legacies of the unlawful or biased policing practices on which they are built.

One of the most interesting aspects of the report is that it shows that the problem of systemic bias and dirty data is not only difficult but insurmountable (Richardson et al. 2019). I believe that the greatest and most risky illusion in the AI industry is that companies overestimate how systemic bias could be corrected. The problem is that when it comes to human profiling, algorithms are always fallacious. This implies that they cannot be corrected, rectified, or fixed. In addition, data will always be biased because it will always be the product of the social context that produced it, of its power relationships and human fears, beliefs, and desires.

When we think about the future risks of AI technologies, we need to seriously start tackling questions about human error. We need to recognize that AI technologies can be very helpful to fight climate change or specific diseases, but we cannot use these tools to profile human beings without recognizing that these technologies are always biased, unexplainable, and inaccurate when it comes to understanding humans. Human error in algorithms cannot be corrected or fixed.

AI FOR GOOD?

Once we realize that the human error cannot be fixed and that AI technologies will always be biased and inaccurate when it comes to human profiling, there is another important step that we should take if we want to protect our democratic futures. When it comes to human profiling, "Good AI" does not exist. During the Child | Data | Citizen project, I stumbled upon the example of Fama Technologies. In 2019, CNN reported that the company was helping businesses to identify risky applicants before they got hired by using artificial intelligence technologies that analyzed a person's public digital footprint and found problematic behaviors like sexual harassment, bigotry, and bullying (Vasel 2019). When I read the article, I immediately thought that, of course an AI technology able to highlight problematic behaviors such as sexism, bigotry, and bullying in the workplace should be very welcome.

So I went onto their website to find out more, and realized how problematic good intentions could be when it came to AI design and business models. The company was founded in 2015, and in 2018 they screened more than 20 million candidates. They found that 14.2 percent of the people they scanned had flags for misogyny and sexism; 10.3 percent had flags for bigotry; 11.7 percent had flags for violence, drugs, and crime; and 24.9 percent had flags for "*company-specific issues or threats*."

As I looked at their website (figure 9.1) and their statistics, my personal red flags as a data ethnographer were immediately triggered: How could they be sure about a person's intention simply by looking at the digital footprint? How could candidates challenge these profiles if they were inaccurate or wrong? My worry was validated when, in January 2020, I stumbled upon a tweet by @kmlefranc, in which she described her dismay when she had to get a background check for her job. The report was a 300+ PDF document with all the tweets that she liked that had the word *f**k* in it.

To protect itself against libel, the company has been very careful in explaining that they do not score people or recommend candidates for hiring, what they do is to present employers with distilled data so that they can make their own decisions (Ha 2015). Data is never distilled or raw; it is gathered through

Figure 9.1
Fama Technologies website.

practices of framing and judgment. Hence although the company policy may protect them from legal libels, the simple fact that they find flags in people's digital footprints implies that their AI is in fact judging people.

Yet what really worried me by looking at their website was that a technology that seemed to have been designed mostly for the good of society, that is, to exclude sexism, misogyny, bigotry, and other forms of toxic behavior from the workplace, could actually be customized. In fact, Fama Technologies promotes itself for being able to customize its technologies to fit their costumers' hiring standards, code of conduct, and employment policies. When I read their promotional blurb I wondered whether the company had a code of ethics of a sort, like a list of specific clients, codes of conduct, or employment policies that they would not agree to. I could not find an answer on their website. Yet I believe that AI companies really need to start asking and addressing more questions about the ways in which their technologies, which are designed for the good, can actually be customized to reinforce existing or new inequalities.

OUR DATA FUTURES

When we think about the impact of AI, we need to remember that data can always be manipulated and turned against us. Jill, one of the mothers that I encountered in this life-changing research journey, once told me

> J: You really don't know what the future holds; we would have never predicted that they would use social media data to manipulate people and sway their votes, but here we go. … In the future there will be more AI and more connected data, and the problem is that we don't know what will happen because we have signed everything away by agreeing to terms and conditions.

Like Jill, many of the parents I worked with understood that data and AI systems could turn against them and their children, and they worried about their children's data futures. As shown, however, many felt alone in the struggle of protecting their children's privacy; they felt powerless and unable to make a difference.

Parents are right to worry about their children's data futures. The main problem that we face today is that we are trusting technologies to make key decisions about individual lives on the basis of data traces, but these technologies will always be inaccurate and biased. Algorithmic error and bias may be reduced but it can never be eradicated. Companies cannot fix the problem, and individuals can't really protect themselves. This is a systemic problem, and the only solution to this has to be a political solution. Governments must step up to protect our rights, and they need to act now to change current regulations and modify current cultural practices. We need to recognize that we cannot fix the human error in algorithms. If we do not understand this, we are not going to be able to protect ourselves and our children or their futures.

When I think about the AI and data-driven futures that we are building for our kids, I am scared. I worry that my children will be exposed to all sorts of algorithmic bias and error. I also worry that they are going to be judged by the data traces that are collected about them today, and that this will prevent them from becoming their own persons. It is time for us to do something before it's too late. Our fight against the injustice of our data and AI-driven economies, however, cannot and should not be an individual pursuit. We need to get together as institutions, organizations, and as a collective entity and shift current debates from individual responsibility and privacy to a collective demand to protect our right to freedom of expression, self-representation, and nondiscrimination. In other words, we need to fight for our right to data justice. Until we challenge the techno-fetishism and data-driven solutions of surveillance capitalism, until we recognize that there is something fundamentally unjust and wrong in the system as a whole, we will be unable to protect ourselves and our children or hope for a more just future.

NOTES

INTRODUCTION

1. All names in the book are fictional to protect the anonymity of the interviewees.

2. A note on ethics: Although it was impossible to seek informed consent on all the different occasions and from all the different parents that I observed, I always made sure that the parents I had contact with were aware of my research and that if something emerged from an informal conversation or observation I would have their permission to add that to my notes. I treated all the information as completely anonymous.

3. The term revolution has been used by a variety of authors from Mayer-Schönberger and Cuckier (2013) to Kitchin (2014) and Lohr (2015) just to mention a few.

CHAPTER 4: EDUCATIONAL DATA

1. GL Assessment | GL Education. 2019. https://www.gl-assessment.co.uk/.

2. In the US in 2011—the same year as the amendment to FERPA was proposed—the Shared Learning Collaborative was created, which in 2013 became inBloom. Funded by the Gates and Carnegie Foundations, inBloom aimed to become a key educational data infrastructure in the US by collecting student names, addresses, grades, test scores, social class, ethnicity, special education status, disciplinary status, and other sensitive and context-specific personal data. In the high of the Snowden affairs and the news of multiple data breaches affecting different companies, inBloom suffered grassroots criticism, as it transpired that the company planned to share educational data with for-profit vendors. Following the public backlash at the national level, the inBloom initiative ended in 2014 (Bulger et al. 2017). The inBloom case was perhaps the most well-known (and failed) example in the US of a company who admitted to be sharing educational data with third parties for profit. Yet currently, even when educational technology companies do not share educational data for profit, the market in the US relies on third-party disclosures and data sharing agreements.

CHAPTER 5: HOME LIFE DATA

1. In chapter 2, I discussed it is important to understand the emotional appeal of data tracking in family life by challenging dismissive claims of data fetishism, but

we must acknowledge that the age of surveillance capitalism, businesses, and organizations are constantly reinforcing forms of data fetishism.

CHAPTER 6: SOCIAL MEDIA DATA

1. Google is also directly gathering children's data through the Family Link Parental Control account, which enables parents to add their children to their Google account. What is particularly fascinating about the Family Link account is that Google explicitly lets parents know that when children turn 13 (or applicable age depending on country) they can "Manage their Google account." Yet there is no mention as to what happens to all the data collected from them before they turned 13. Is that data integrated into a Unique ID profile?

REFERENCES

Aldrich, Francis. 2006. "Smart Homes: Past, Present and Future." In *Inside the Smart Home*, edited by Richard Harper, 17–41. Springer Science & Business Media.

Alleyne, Brian. 2000. "Personal Narrative & Activism: A Bio-Ethnography of 'Life Experience with Britain.'" PhD thesis, Goldsmiths College.

Andersson, Hilary. 2018. "Social Media Is 'Deliberately' Addictive." *BBC*, July 4, 2018, Technology. https://www.bbc.com/news/technology-44640959.

Andreassen, Cecilie, Joël Billieux, Mark D. Griffiths, Daria J. Kuss, Zsolt Demetrovics, Elvis Mazzoni, and Ståle Pallesen. 2016. "The Relationship between Addictive Use of Social Media and Video Games and Symptoms of Psychiatric Disorders: A Large-Scale Cross-Sectional Study." *Psychology of Addictive Behaviors* 30, no. 2: 252–62. https://doi.org/10.1037/adb0000160.

Andrejevic, Mark. 2004. "The Work of Watching One Another: Lateral Surveillance, Risk, and Governance." *Surveillance & Society* 2, no. 4: 479–497. https://doi.org/10.24908/ss.v2i4.3359.

Andrejevic, Mark, and Neil Selwyn. 2020. "Facial Recognition Technology in Schools: Critical Questions and Concerns." *Learning, Media and Technology* 45, no. 2: 115–128. https://doi.org/10.1080/17439884.2020.1686014.

Angwin, Julia, Ariana Tobin, and Marlene Varner. 2017. "Facebook (Still) Letting Housing Advertisers Exclude Users by Race." *ProPublica*, November 21, 2017. https://www.propublica.org/article/facebook-advertising-discrimination-housing-race-sex-national-origin.

Arora, Payal. 2016. "Bottom of the Data Pyramid: Big Data and the Global South." *International Journal of Communication* 10, no. 0: 19.

Arora, Payal. 2019. "Decolonizing Privacy Studies." *Television & New Media* 20, no. 4: 366–78. https://doi.org/10.1177/1527476418806092.

BabyCenter LLC. 2019. "BabyCenter Privacy Policy." BabyCenter. https://www.babycenter.com/help-privacy. Accessed May 5, 2017.

BabyCentre UK. 2019. "Privacy Policy." BabyCentre. 2019. Accessed May 5, 2017. https://www.babycentre.co.uk/e7814/privacy-policy.

Barassi, Veronica. 2015. *Activism on the Web: Everyday Struggles against Digital Capitalism*. Routledge.

Barassi, Veronica. 2016. "Datafied Citizens? Social Media Activism, Digital Traces and the Question about Political Profiling." *Communication and the Public*, December. https://doi.org/10.1177/2057047316683200.

Barassi, Veronica. 2017a. "BabyVeillance? Expecting Parents, Online Surveillance and the Cultural Specificity of Pregnancy Apps." *Social Media + Society* 3, no. 2: 2056305117707188. https://doi.org/10.1177/2056305117707188.

Barassi, Veronica. 2017b. "Digital Citizens? Data Traces and Family Life." *Contemporary Social Science* 12, no. 1–2: 84–95. https://doi.org/10.1080/21582041.2017.1338353.

Barassi, Veronica. 2018. "Home Life Data and Children's Privacy." Call for Evidence Submission Information, Commissioner's Office. Goldsmiths, University of London.

Barassi, Veronica. 2019. "Datafied Citizens in the Age of Coerced Digital Participation:" *Sociological Research Online*, June. https://doi.org/10.1177/1360780419857734.

Barassi, Veronica, and Patricia Scanlon. 2019. "Voice Prints and Children's Rights." Goldsmiths, University of London. http://childdatacitizen.com/voice-prints-childrens-rights/.

Barocas, Solon, and Andrew D. Selbst. 2016. "Big Data's Disparate Impact." *104 California Law Review* 671. Available at SSRN: https://papers.ssrn.com/abstract=2477899.

Beer, David. 2019. *The Data Gaze: Capitalism, Power and Perception*. Sage.

Bellamy-Foster, John, and Robert McChesney. 2014. "Monthly Review | Surveillance Capitalism." *Monthly Review* (blog). July 1, 2014. https://monthlyreview.org/2014/07/01/surveillance-capitalism/.

Berleur, Jacques, Markku I. Nurminen, and John Impagliazzo. 2007. "Social Informatics: An Information Society for All?" In *Remembrance of Rob Kling: Proceedings of the Seventh International Conference "Human Choice and Computers"* (HCC7), IFIP TC 9, Maribor, Slovenia, September 21–23, 2006. Springer.

Besnier, Niko. 2009. *Gossip and the Everyday Production of Politics*. University of Hawai'i Press.

Bessant, Claire. 2017. "Parental Rights to Publish Family Photographs versus Children's Rights to a Private Life." *Entertainment Law Review* 28 (April): 43–46.

Best, D. 2006. "Web 2.0: Next Big Thing or Next Big Internet Bubble?" Technische Universiteit Eindhoven.

Blum-Ross, Alicia, and Sonia Livingstone. 2017. "'Sharenting,' Parent Blogging, and the Boundaries of the Digital Self." *Popular Communication* 15, no. 2: 110–125. https://doi.org/10.1080/15405702.2016.1223300.

Boellstorff, Tom. 2013. "Making Big Data, in Theory." *First Monday* 18, no. 10. http://firstmonday.org/ojs/index.php/fm/article/view/4869.

Bory, Paolo. 2019. "Deep New: The Shifting Narratives of Artificial intelligence from Deep Blue to AlphaGo." *Convergence* 25, no. 4 (February): 627–642. https://doi.org/10.1177/1354856519829679.

Bourdieu, Pierre. 1970. "The Berber House or the World Reversed." *Social Science Information* 9, no. 2: 151–170. https://doi.org/10.1177/053901847000900213.

Bowles, Nellie. 2019. "Silicon Valley Came to Kansas Schools: That Started a Rebellion." *New York Times*, April 25, 2019, Technology. https://www.nytimes.com/2019/04/21/technology/silicon-valley-kansas-schools.html.

boyd, danah, and Kate Crawford. 2012. "Critical Questions for Big Data." *Information, Communication & Society* 15, no. 5: 662–679. https://doi.org/10.1080/1369118X.2012.678878.

Bradbury, Alice, and Guy Roberts-Holmes. 2016. "'They Are Children … Not Robots, Not Machines': The Introduction of Reception Baseline Assessment." UCL Institute of Education. http://www.betterwithoutbaseline.org.uk/uploads/2/0/3/8/20381265/baseline_assessment_2.2.16-_10404.pdf.

Brockell, Gillian. 2018. "Perspective | Dear Tech Companies, I Don't Want to See Pregnancy Ads after My Child Was Stillborn." *Washington Post*, December 12, 2018, On Parenting: Perspective. https://www.washingtonpost.com/lifestyle/2018/12/12/dear-tech-companies-i-dont-want-see-pregnancy-ads-after-my-child-was-stillborn/.

Broussard, Meredith. 2018. *Artificial Unintelligence: How Computers Misunderstand the World*. The MIT Press.

Bucher, Taina. 2017. "The Algorithmic Imaginary: Exploring the Ordinary Affects of Facebook Algorithms." *Information, Communication & Society* 20 (1): 30–44. https://doi.org/10.1080/1369118X.2016.1154086.

Bulger, Monica, Patrick McCormick, and Mikaela Pitcan. 2017. "The Legacy of InBloom." Data and Society Report.

Bullock, William, Liang Xu, and Li Zhou. n.d. Predicting Household Demographics based on Image Data. US Patent, filed November 15, 2018. Accessed July 29, 2019. http://appft.uspto.gov/netacgi/nph-Parser?Sect1=PTO1&Sect2=HITOFF&d=PG01&p=1&u=%2Fnetahtml%2FPTO%2Fsrchnum.html&r=1&f=G&l=50&s1=%2220180332140%22.PGNR.&OS=DN/20180332140&RS=DN/20180332140.

Business Wire. 2010. "Digital Birth: Welcome to the Online World." *Business Wire*, October 6, 2010. https://www.businesswire.com/news/home/20101006006722/en/Digital-Birth-Online-World.

Campbell-Dollaghan, Kelsey. 2015. "Artificial Neural Networks Can Day Dream—Here's What They See." *Gizmodo*, June 18, 2015. https://gizmodo.com/these-are-the-incredible-day-dreams-of-artificial-neura-1712226908.

Casinelli, Matthew. 2018. "Apple's Shortcuts Will Flip the Switch on Siri's Potential." *TechCrunch* (blog). July 9, 2018. Accessed December 12, 2018. http://social.techcrunch.com/2018/07/08/shortcuts-will-flip-the-switch-for-apple-on-siris-potential-just-not-how-you-think/.

Castells, Manuel. 1996. *The Rise of the Network Society: Information Age: Economy, Society, and Culture v. 1*. Wiley-Blackwell.

Castells, Manuel. 1997. *The Power of Identity: The Information Age—Economy, Society, and Culture: 2*. Wiley-Blackwell.

Castells, Manuel. 2007. "Communication, Power and Counter-Power in the Network Society." *International Journal of Communication* 1, no. 1: 29.

Castells, Manuel. 2009. *Communication Power*. Oxford University Press.

CB Insights. 2018. "How Google Plans to Use AI to Reinvent the $3 Trillion US Healthcare Industry." CB Information Services.

Chakravartty, Paula, Rachel Kuo, Victoria Grubbs, and Charlton McIlwain. 2018. "#CommunicationSoWhite." *Journal of Communication* 68, no. 2: 254–266. https://doi.org/10.1093/joc/jqy003.

Chambers, Deborah. 2016. *Changing Media, Homes and Households: Cultures, Technologies and Meanings*. Routledge.

Chantler, Cyril, Trevor Clarke, and Richard Granger. 2006. "Information Technology in the English National Health Service." *JAMA* 296, no. 18: 2255–2258. https://doi.org/10.1001/jama.296.18.2255.

Chavez, Tom, Chris O'Hara, and Vivek Vaidya. 2018. *Data Driven: Harnessing Data and AI to Reinvent Customer Engagement*. McGraw-Hill Education.

Cheney-Lippold, John. 2017. *We Are Data: Algorithms and The Making of Our Digital Selves*. NYU Press.

Citron, Danielle Keats. 2007. "Technological Due Process." *Washington University Law Review* 85: 1249–1313. Available at SSRN: https://papers.ssrn.com/abstract=1012360.

Citron, Danielle Keats, and Frank A. Pasquale. 2014. "The Scored Society: Due Process for Automated Predictions." Available at SSRN: https://papers.ssrn.com/abstract=2376209.

Clifford, James, and George E. Marcus, eds. 2010. *Writing Culture: The Poetics and Politics of Ethnography*. 2nd ed., 25th anniversary ed. University of California Press.

Coady, Margaret. 2009. "Being and Becomings: Historical and Philosophical Considerations of the Child as Citizen." In *Young Children as Active Citizens: Principles, Policies and Pedagogies*, edited by Glenda Mac Naughton, Patrick Hughes, and Kylie Smith, 3–16. Cambridge Scholars Publishing.

Cohen, Anthony. 1994. *Self Consciousness: An Alternative Anthropology of Identity*. Routledge.

Cole, David. 2019. "The Chinese Room Argument." In *The Stanford Encyclopedia of Philosophy*, edited by Edward N. Zalta. Spring 2019. Metaphysics Research Lab, Stanford University. https://plato.stanford.edu/archives/spr2019/entries/chinese-room/.

Constine, Josh. 2018. "Facebook Rolls out AI to Detect Suicidal Posts before They're Reported." *TechCrunch*. http://social.techcrunch.com/2017/11/27/facebook-ai-suicide-prevention/.

Cookiebot. 2019. "Cookiebot Report: Hidden Tracking of Citizens on EU Government and Health Sector Websites." Cookiebot. https://www.cookiebot.com/en/cookiebot-report/.

Costanza-Chock, Sasha. 2018. "Design Justice, A.I., and Escape from the Matrix of Domination." *Journal of Design and Science* (July). https://doi.org/10.21428/96c8d426.

Couldry, Nick, and Ulises A. Mejias. 2019. "Data Colonialism: Rethinking Big Data's Relation to the Contemporary Subject." *Television & New Media* 20, no. 4: 336–349. https://doi.org/10.1177/1527476418796632.

Couldry, Nick, Hilde Stephansen, Aristea Fotopoulou, Richard MacDonald, Wilma Clark, and Luke Dickens. 2014. "Digital Citizenship? Narrative Exchange and the Changing Terms of Civic Culture." *Citizenship Studies* 18, no. 6–7: 615–629. https://doi.org/10.1080/13621025.2013.865903.

Crawford, Kate, and Ryan Calo. 2016. "There Is a Blind Spot in AI Research." *Nature News* 538, no. 7625: 311. https://doi.org/10.1038/538311a.

Crawford, Kate, and Jason Schultz. 2014. "Big Data and Due Process: Toward a Framework to Redress Predictive Privacy Harms." *Boston College Law Review* 55, no. 1: 93.

Crenshaw, Kimberle. 1989. "Demarginalizing the Intersection of Race and Sex: A Black Feminist Critique of Antidiscrimination Doctrine, Feminist Theory and Antiracist Politics." *University of Chicago Legal Forum* 1989, no. 1: 139–67.

Dahlgren, Peter. 2009. *Media and Political Engagement: Citizens, Communication and Democracy*. Cambridge University Press.

Davies, Harry. 2015. "Ted Cruz Campaign Using Firm That Harvested Data on Millions of Unwitting Facebook Users." *The Guardian*, December 11, 2015, US news. https://www.theguardian.com/us-news/2015/dec/11/senator-ted-cruz-president-campaign-facebook-user-data.

Day, Matt, Giles Turner, and Natalia Drozdiak. 2019. "Thousands of Amazon Workers Listen to Alexa Users' Conversations." *Time*, November 4, 2019. https://time.com/5568815/amazon-workers-listen-to-alexa/.

Defend Digital Me. 2019. "State of Data 2019 Report." Defend Digital Me Campaign. https://defenddigitalme.com/.

Dencik, Lina, Arne Hintz, and Jonathan Cable. 2016. "Towards Data Justice? The Ambiguity of Anti-Surveillance Resistance in Political Activism." *Big Data & Society* 3, no. 2. https://doi.org/10.1177/2053951716679678.

Dencik, Lina, Arne Hintz, and Zoe Carey. 2017. "Prediction, Pre-Emption and Limits to Dissent: Social Media and Big Data Uses for Policing Protests in the United Kingdom:" *New Media & Society* 20, no. 4: 1433–1450. https://doi.org/10.1177/1461444817697722.

Dencik, Lina, Arne Hintz, Joanna Redden, and Emiliano Treré. 2019. "Exploring Data Justice: Conceptions, Applications and Directions." *Information, Communication & Society* 22, no. 7: 873–881. https://doi.org/10.1080/1369118X.2019.1606268.

Draper, Nora A, and Joseph Turow. 2019. "The Corporate Cultivation of Digital Resignation." *New Media & Society* 21, no. 8:1824–1839. https://doi.org/10.1177/1461444819833331.

Earls, Felton J. 2011. *The Child as Citizen*. Sage.

Eisenstat, Yael. 2019. "The Real Reason Tech Struggles with Algorithmic Bias." *Wired*, February 12, 2019. https://www.wired.com/story/the-real-reason-tech-struggles-with-algorithmic-bias/.

Elgan, Mike. 2018. "The Case against Teaching Kids to Be Polite to Alexa." *Fast Company*, June 24, 2018. https://www.fastcompany.com/40588020/the-case-against-teaching-kids-to-be-polite-to-alexa.

Elmer, Greg. 2004. *Profiling Machines: Mapping the Personal Information Economy*. The MIT Press.

Elmer, Greg, and Andy Opel. 2008. *Preempting Dissent: The Politics of an Inevitable Future*. Arbeiter Ring.

Escobar, Arturo. 2004. "Identity." In *A Companion to the Anthropology of Politics,* edited by David Nugent and Joan Vincent, 248–267. Blackwell.

Eubanks, Virginia. 2014. "Want to Predict the Future of Surveillance? Ask Poor Communities." *The American Prospect*, January 15, 2014. https://prospect.org/article/want-predict-future-surveillance-ask-poor-communities.

Eubanks, Virginia. 2018. *Automating Inequality: How High-Tech Tools Profile, Police, and Punish the Poor*. St. Martin's Press.

European Commission. 2017. "Questions and Answers—Data Protection Reform Package." http://europa.eu/rapid/press-release_MEMO-17-1441_en.htm.

Evans, R. S. 2016. "Electronic Health Records: Then, Now, and in the Future." *Yearbook of Medical Informatics*, Suppl 1 (May): S48–S61. https://doi.org/10.15265/IYS-2016-s006.

Facebook Inc. 2017. "Facebook Messenger Kids Data Policy." April 12, 2017. https://www.facebook.com/legal/messengerkids/privacypolicy.

Facebook Inc. 2018a. "Data Policy." Facebook. April 19, 2018. https://www.facebook.com/privacy/explanation.

Facebook Inc. 2018b. "Terms of Service." April 19, 2018. https://www.facebook.com/legal/terms.

Farr, Christina. 2018. "Facebook Sent a Doctor on a Secret Mission to Ask Hospitals to Share Patient Data." *CNBC*. https://www.cnbc.com/2018/04/05/facebook-building-8-explored-data-sharing-agreement-with-hospitals.html.

Federal Trade Commission. 2014. "Data Brokers: A Call for Transparency and Accountability." https://www.ftc.gov/system/files/documents/reports/data-brokers-call-transparency-accountability-report-federal-trade-commission-may-2014/140527databrokerreport.pdf.

Fenton, Natalie, and Veronica Barassi. 2011. "Alternative Media and Social Networking Sites: The Politics of Individuation and Political Participation." *Communication Review* 14, no. 3: 179–196. https://doi.org/10.1080/10714421.2011.597245.

Ferguson, Christopher J. 2018. "Debunking the 6 Biggest Myths about 'Technology Addiction.'" *The Conversation*. http://theconversation.com/debunking-the-6-biggest-myths-about-technology-addiction-95850.

Floridi, Luciano, ed. 2015. *The Onlife Manifesto: Being Human in a Hyperconnected Era*. Springer. https://www.springer.com/gp/book/9783319040929.

Foucault, Michel. 2012. *Discipline & Punish: The Birth of the Prison*. Knopf Doubleday.

Friedman, Batya, and Helen Nissenbaum. 1996. "Bias in Computer Systems." *ACM Transactions on Information Systems* 14, no. 3: 330–347. https://doi.org/10.1145/230538.230561.

Fuchs, Christian. 2013. "Digital Prosumption Labour on Social Media in the Context of the Capitalist Regime of Time." *Time & Society* 23, no. 1: 97–123. https://doi.org/10.1177/0961463X13502117.

Gangadharan, Seeta Peña. 2012. "Digital Inclusion and Data Profiling." *First Monday* 17, no. 5. http://firstmonday.org/ojs/index.php/fm/article/view/3821.

Gangadharan, Seeta Peña. 2015. "The Downside of Digital Inclusion: Expectations and Experiences of Privacy and Surveillance among Marginal Internet Users." *New Media & Society* 18, no. 4: 597–615. https://doi.org/10.1177/1461444815614053.

Gangadharan, Seeta Peña, and Jędrzej Niklas. 2019. "Decentering Technology in Discourse on Discrimination." *Information, Communication & Society* 22, no. 7: 882–899. https://doi.org/10.1080/1369118X.2019.1593484.

Gayle, Damien. 2016. "Pupil Data Shared with Home Office to 'create Hostile Environment' for Illegal Migrants." *The Guardian*, December 15, 2016, sec. UK news.

https://www.theguardian.com/uk-news/2016/dec/15/pupil-data-shared-with-home-office-to-identify-illegal-migrants.

Giddens, Anthony. 1991. *The Consequences of Modernity*. Stanford University Press.

Gill, Rosalind, and Andy Pratt. 2008. "In the Social Factory? Immaterial Labour, Precariousness and Cultural Work." *Theory, Culture & Society* 25, no. 7–8: 1–30. https://doi.org/10.1177/0263276408097794.

Gillum, Jack, and Ariana Tobin. 2019. "Facebook Won't Let Employers, Landlords or Lenders Discriminate in Ads Anymore." *ProPublica*, March 19, 2019. https://www.propublica.org/article/facebook-ads-discrimination-settlement-housing-employment-credit.

Gilman, Michele E. 2012. "The Class Differential in Privacy Law." *Brooklyn Law Review* 77, no. 4: 1389–1444.

Gitelman, Lisa. 2013. *"Raw Data" Is an Oxymoron*. The MIT Press.

Google LCC. 2017. "Google Terms of Service—Privacy & Terms—Google." October 15, 2017. https://policies.google.com/terms?hl=en-US.

Google LCC. 2019a. "Privacy Policy—Privacy & Terms—Google." January 22, 2019. https://policies.google.com/privacy?hl=en-US#infocollect.

Google LCC. 2019b. "YouTube Kids Privacy Notice—YouTube." January 31, 2019. https://kids.youtube.com/t/privacynotice.

Gould, Stephen Jay. 2006. *The Mismeasure of Man (Revised & Expanded)*. W. W. Norton.

Graeber, David. 2007. *Possibilities: Essays on Hierarchy, Rebellion, and Desire*. AK Press.

Grand View Research. 2018. "MHealth Market Size, Share | Global Industry Trends Report, 2018–2025." Market Trends. Grand View Research. Report ID: 978-1-68038-076-7. https://www.grandviewresearch.com/industry-analysis/mhealth-market.

Greengard, Samuel. 2015. *The Internet of Things*. The MIT Press.

Gregg, Melissa. 2011. *Work's Intimacy*. Polity Press.

Grundy, Quinn, Kellia Chiu, Fabian Held, Andrea Continella, Lisa Bero, and Ralph Holz. 2019. "Data Sharing Practices of Medicines Related Apps and the Mobile Ecosystem: Traffic, Content, and Network Analysis." *BMJ* 364 (March): l920. https://doi.org/10.1136/bmj.l920.

Ha, Anthony. 2015. "Fama Helps Businesses Find Social Media 'Red Flags' Before Hiring Someone." *TechCrunch* (blog). 2015. http://social.techcrunch.com/2015/09/10/fama-technologies/.

Hamilton, Isobel Asher. 2018. "There's Zero Compelling Evidence Showing Tech Is as Addictive as Cocaine, According to Scientists." *Business Insider*, October 7, 2018. https://www.businessinsider.com/theres-no-evidence-that-tech-is-as-addictive-as-cocaine-2018-7.

Hargittai, Eszter, and Alice Marwick. 2016. "'What Can I Really Do?' Explaining the Privacy Paradox with Online Apathy." *International Journal of Communication* 10: 21.

Harvard Law Today. 2019. "At Harvard Law, Zittrain and Zuckerberg Discuss Encryption, 'Information Fiduciaries' and Targeted Advertisements." *Harvard Law Today*. February 20, 2019. https://today.law.harvard.edu/at-harvard-law-zittrain-and-zuckerberg-discuss-encryption-information-fiduciaries-and-targeted-advertisements/.

Harvey, David. 1991. *The Condition of Postmodernity: An Enquiry into the Origins of Cultural Change*. Wiley-Blackwell.

Harvey, David. 2003. "The Fetish of Technology: Causes and Consequences." *Macalester International* 13, no. 1: 1–30. http://digitalcommons.macalester.edu/macintl/vol13/iss1/7.

Harwell, Drew. 2018. "Wanted: The 'Perfect Babysitter.' Must Pass AI Scan for Respect and Attitude." *Washington Post*, November 23, 2018, The Switch. https://www.washingtonpost.com/technology/2018/11/16/wanted-perfect-babysitter-must-pass-ai-scan-respect-attitude/.

Hassan, Robert. 2003. "Network Time and the New Knowledge Epoch." *Time & Society* 12, no. 2–3: 225–241. https://doi.org/10.1177/0961463X030122004.

Hassan, Robert. 2007. "Network Time." In *24/7: Time and Temporality in the Network Society*, edited by Robert Hassan and Roland Purser, 37–62. Stanford Business Books.

Hassan, Robert. 2009. *Empires of Speed: Time and the Acceleration of Politics and Society*. Brill.

HealthIT. n.d. "What Is an Electronic Health Record (EHR)?" HealthIT.gov. Accessed July 29, 2019. https://www.healthit.gov/faq/what-electronic-health-record-ehr.

Heather, Ben. 2018. "EMIS Moving 40 Million UK Patient Records to Amazon Web Services." *Thought Leadership In Digital Health*. https://www.healthcare.digital/single-post/2018/12/02/EMIS-moving-40-million-UK-patient-records-to-Amazon-Web-Services.

Heeks, Richard, and Jaco Renken. 2018. "Data Justice for Development: What Would It Mean?" *Information Development* 34, no. 1: 90–102. https://doi.org/10.1177/0266666916678282.

Hess, Robert D., Judith V. Torney-Purta, and Jaan Valsiner. 2005. *The Development of Political Attitudes in Children*. Aldine Transaction.

Hildebrandt, Mireille, and Serge Gutwirth. 2008. *Profiling the European Citizen: Cross-Disciplinary Perspectives*. Springer Science & Business Media.

Hintz, Arne, Lina Dencik, and Karin Wahl-Jorgensen. 2017. "Digital Citizenship and Surveillance| Digital Citizenship and Surveillance Society—Introduction." *International Journal of Communication* 11: 9.

Hintz, Arne, Lina Dencik, and Karin Wahl-Jorgensen. 2018. *Digital Citizenship in a Datafied Society*. Polity Press.

Hoffmann, Anna Lauren. 2019. "Where Fairness Fails: Data, Algorithms, and the Limits of Antidiscrimination Discourse." *Information, Communication & Society* 22, no. 7: 900–915. https://doi.org/10.1080/1369118X.2019.1573912.

Hornborg, Alf. 1992. "Machine Fetishism, Value, and the Image of Unlimited Good: Towards a Thermodynamics of Imperialism." *Man* 27, no. 1: 1–18. https://doi.org/10.2307/2803592.

Househ, Mowafa, Elizabeth Borycki, and Andre Kushniruk. 2014. "Empowering Patients through Social Media: The Benefits and Challenges." *Health Informatics Journal* 20, no. 1: 50–58. https://doi.org/10.1177/1460458213476969.

HSCIC (Health and Social Care Information Centre). 2015. "Information and Technology for Better Care. Health and Social Care Information Centre Strategy 2015–2020." NHS Digital. 2015. https://digital.nhs.uk/about-nhs-digital/corporate-information-and-documents/our-strategy.

Humphreys, Lee. 2018. *The Qualified Self: Social Media and the Accounting of Everyday Life*. The MIT Press.

ICO (Information Commissioner's Office). 2019. "Update Report into AdTech and Real Time Bidding." https://ico.org.uk/media/about-the-ico/documents/2615156/adtech-real-time-bidding-report-201906.pdf.

Isin, Engin, and E. Ruppert. 2015. *Being Digital Citizens*. Rowman & Littlefield International.

Ito, Mizuko. 2012. *Engineering Play: A Cultural History of Children's Software*. The MIT Press.

Jin, Huafeng, and Shuo Wang. 2018. Voice-based determination of physical and emotional characteristics of users. US Patent 10096319, filed March 13, 2017, and issued October 9, 2018. http://patft.uspto.gov/netacgi/nph-Parser?Sect2=PTO1&Sect2=HITOFF&p=1&u=%2Fnetahtml%2FPTO%2Fsearch-bool.html&r=1&f=G&l=50&d=PALL&RefSrch=yes&Query=PN%2F10096319.

Juniper Research. 2018. "Digital Voice Assistants Platforms Research and Forecasts." https://www.juniperresearch.com/researchstore/innovation-disruption/digital-voice-assistants/platforms-revenues-opportunities?utm_source=juniperpr&utm_campaign=digitalvoiceassistantspr1&utm_medium=email.

Kafka, Franz. (1925) 2020. *The Trial*. Pan Macmillan.

Kahn, Jeremy, and John Lauerman. 2018. "Google Taking Over Health Records Raises Patient Privacy Fears." *Bloomberg Technology*, November 20, 2018. https://www.bloomberg.com/news/articles/2018-11-21/google-taking-over-health-records-raises-patient-privacy-fears.

Kelion, Leo. 2019. "Amazon Sued over Alexa Child Recordings." *BBC News*, June 13, 2019, Technology. https://www.bbc.com/news/technology-48623914.

Khan, Shehab. 2016. "Teenager Sues Parents over Embarrassing Childhood Facebook Pictures." *The Independent*, September 14, 2016. http://www.independent.co.uk/news/world/europe/teenager-sues-parents-over-embarrassing-childhood-pictures-on-facebook-austria-a7307561.html.

Kharpal, Arjun. 2017. "Stephen Hawking Says A.I. Could Be 'Worst Event in the History of Our Civilization.'" *CNBC*. November 6, 2017. https://www.cnbc.com/2017/11/06/stephen-hawking-ai-could-be-worst-event-in-civilization.html.

King, Pippa. 2018. "Biometrics in Schools: Consent Needed by Schools to Process Biometric Data." *Biometrics in Schools* (blog). August 30, 2018. https://pippaking.blogspot.com/2018/08/consent-needed-by-schools-to-process.html.

Kinsella, Brett. 2018. "Amazon Alexa Skill Count Surpasses 30,000 in the US" *Voicebot*, March 22, 2018. https://voicebot.ai/2018/03/22/amazon-alexa-skill-count-surpasses-30000-u-s/.

Kitchin, Rob. 2014. *The Data Revolution: Big Data, Open Data, Data Infrastructures and Their Consequences*. Sage.

Kleinberg, Sarah. 2018. "5 Ways Voice Assistance Is Reshaping Consumer Behavior." Think with Google. 2018. https://www.thinkwithgoogle.com/consumer-insights/voice-assistance-consumer-experience/.

Kuss, Daria Joanna, and Mark D. Griffiths. 2012. "Internet Gaming Addiction: A Systematic Review of Empirical Research." *International Journal of Mental Health and Addiction* 10, no. 2: 278–296. https://doi.org/10.1007/s11469-011-9318-5.

Latour, Bruno, and Steve Woolgar. 1986. *Laboratory Life: The Construction of Scientific Facts*. Edited by Jonas Salk. Reprint edition. Princeton University Press.

Lawn, Martin. 2013. "Voyages of Measurement in Education in the Twentieth Century: Experts, Tools and Centres." *European Educational Research Journal* 12, no. 1: 108–119. https://doi.org/10.2304/eerj.2013.12.1.108.

Leaver, Tama. 2015. "Born Digital? Presence, Privacy, and Intimate Surveillance." In *Re-Orientation: Translingual Transcultural Transmedia: Studies in Narrative, Language, Identity, and Knowledge*, edited by John Hartley and W. Qu, 149–160. Fudan University Press.

Li, Abner. 2019. "Chromebooks Used by 30M Students & Educators Worldwide w/ 40M Google Classroom Users." *9to5Google*. January 22, 2019. https://9to5google.com/2019/01/22/30m-chromebook-education-users/.

Libert, Timothy. 2015. "Privacy Implications of Health Information Seeking on the Web." *Communications of the ACM* 58, no. 3: 68–77. https://doi.org/10.1145/2658983.

Liebes, Tamar, and Rivka Ribak. 1992. "The Contribution of Family Culture to Political Participation, Political Outlook, and Its Reproduction." *Communication Research* 19, no. 5: 618–641. https://doi.org/10.1177/009365092019005004.

Lipu, Merike, and Andra Siibak. 2019. "'Take It Down!': Estonian Parents' and Pre-Teens' Opinions and Experiences with Sharenting." *Media International Australia* 170, no. 1: 57–67. https://doi.org/10.1177/1329878X19828366.

Livingstone, Sonia. 2009. *Children and the Internet: Great Expectations, Challenging Realities*. Polity.

Livingstone, Sonia, and Alicia Blum-Ross. 2017. "The Trouble with 'Screen Time Rules.'" *Parenting for a Digital Future* (blog). June 8, 2017. https://blogs.lse.ac.uk/parenting4digitalfuture/2017/06/08/the-trouble-with-screen-time-rules/.

Livingstone, Sonia, and Keely Franklin. 2018. "Families with Young Children and 'Screen Time.'" *Journal of Health Visiting* 6, no. 9: 434–439. https://doi.org/10.12968/johv.2018.6.9.434.

Livingstone, Sonia, and Julian Sefton-Green. 2016. *The Class: Living and Learning in the Digital Age*. NYU Press.

Livingstone, Sonia, and Amanda Third. 2018. "Children: A Special Case for Privacy?" *Intermedia* 46 (July): 18–23.

Lohr, Steve. 2015. *Data-Ism: The Revolution Transforming Decision Making, Consumer Behavior, and Almost Everything Else*. HarperCollins.

Lorenz, Taylor. 2019. "When Kids Google Themselves." *The Atlantic*, February 20, 2019. https://www.theatlantic.com/technology/archive/2019/02/when-kids-realize-their-whole-life-already-online/582916/.

Lupton, Deborah. 2013. *The Social Worlds of the Unborn*. Springer.

Lupton, Deborah. 2016. *The Quantified Self*. John Wiley & Sons.

Lupton, Deborah, and Gareth M. Thomas. 2015. "Playing Pregnancy: The Ludification and Gamification of Expectant Motherhood in Smartphone Apps." *M/C Journal* 18, no. 5. http://journal.media-culture.org.au/index.php/mcjournal/article/view/1012.

Lyon, David. 2001. *Surveillance Society: Monitoring Everyday Life*. McGraw-Hill Education (UK).

Lyon, David. 2007. "Surveillance, Security and Social Sorting: Emerging Research Priorities." *International Criminal Justice Review* 17, no. 3: 161–170. https://doi.org/10.1177/1057567707306643.

MacDonald, Richard. 2018. "A Place for Memory: Family Photo Collections, Social Media and the Imaginative Reconstruction of the Working Class Neighbourhood." In *Picturing the Family: Media, Narrative, Memory*, edited by Silke Arnold-de Simine and Joanna Leal, 151–170. Bloomsbury. https://www.bloomsbury.com/uk/picturing-the-family-9781474283618/.

Madden, Mary, Michele Gilman, Karen Levy, and Alice Marwick. 2017. "Privacy, Poverty, and Big Data: A Matrix of Vulnerabilities for Poor Americans." *Washington University Law Review* 95, no. 1: 53–125.

Marx, Karl. 1990. *Capital: Critique of Political Economy v. 1*. Edited by Ernest Mandel. Translated by Ben Fowkes. New Ed edition. Penguin Classics.

Mauss, Marcel. 1985. "A Category of the Human Mind: The Notion of Person and the Notion of Self." In *The Category of the Person: Anthropology, Philosophy, History*, edited by Michael Carrithers, Steven Collins, and Steven Lukes, 1–26. First paperback edition. Cambridge University Press.

May, Ashley. 2016. "18-Year-Old Sues Parents for Posting Baby Pictures on Facebook." *USA Today*. September 16, 2016. https://www.usatoday.com/story/news/nation-now/2016/09/16/18-year-old-sues-parents-posting-baby-pictures-facebook/90479402/.

Mayer-Schönberger, Viktor, and Kenneth Cukier. 2013. *Big Data: A Revolution That Will Transform How We Live, Work and Think*. John Murray.

Mayer-Schönberger, Viktor, and Kenneth Cukier. 2014. *Learning with Big Data: The Future of Education*. Houghton Mifflin Harcourt.

McCorduck, Pamela. 2004. *Machines Who Think: A Personal Inquiry into the History and Prospects of Artificial intelligence*. 2nd ed. A. K. Peters/CRC Press.

McDaniel, Brandon T., and Jenny S. Radesky. 2018. "Technoference: Parent Distraction with Technology and Associations with Child Behavior Problems." *Child Development* 89, no. 1: 100–109. https://doi.org/10.1111/cdev.12822.

McDonald, Aleecia M., and Lorrie Faith Cranor. 2008. "The Cost of Reading Privacy Policies." *A Journal of Law and Policy for the Information Society* 4, no. 3: 543–568.

McKay, Tom. 2017. "No, Facebook Did Not Panic and Shut Down an AI Program That Was Getting Dangerously Smart." *Gizmodo*. Accessed July 29, 2019. https://gizmodo.com/no-facebook-did-not-panic-and-shut-down-an-ai-program-1797414922.

McQuillan, Dan. 2016. "Algorithmic Paranoia and the Convivial Alternative." *Big Data & Society* 3, no. 2: 2053951716671340. https://doi.org/10.1177/2053951716671340.

Mendelsohn, Stephen. 2012. "US Department of Education Amends Its FERPA Regulations to Allow for Certain Additional Student Disclosures." *National Law Review*, January 2, 2012. https://www.natlawreview.com/article/us-department-education-amends-its-ferpa-regulations-to-allow-certain-additional-student-dis.

Merchant, Brian. 2018. "I Tried Predictim's AI Scan for 'Risky' Babysitters on People I Trust." *Gizmodo*, June 12, 2018. https://gizmodo.com/predictim-claims-its-ai-can-flag-risky-babysitters-so-1830913997.

Metcalfe, Philippa, and Lina Dencik. 2019. "The Politics of Big Borders: Data (in) Justice and the Governance of Refugees." *First Monday* 24, no. 4. https://doi.org/10.5210/fm.v24i4.9934.

Meyer, Robinson. 2018. "The Cambridge Analytica Scandal, in Three Paragraphs." *The Atlantic*, March 20, 2018. https://www.theatlantic.com/technology/archive/2018/03/the-cambridge-analytica-scandal-in-three-paragraphs/556046/.

Milakovich, Michael E. 2012. *Digital Governance: New Technologies for Improving Public Service and Participation*. Routledge.

Milan, Stefania, and Emiliano Treré. 2019. "Big Data from the South(s): Beyond Data Universalism." *Television & New Media* 20, no. 4: 319–335. https://doi.org/10.1177/1527476419837739.

Misra, Satish. 2015. "More than 165,000 Mobile Health Apps Now Available." *IMedicalApps* (blog). September 17, 2015. https://www.imedicalapps.com/2015/09/ims-health-apps-report/.

Moland, Lydia L. 2011. *Hegel on Political Identity: Patriotism, Nationality, Cosmopolitanism*. Northwestern University Press.

Morley, David. 2006. *Media, Modernity, Technology: The Geography of the New*. New Ed edition. Routledge.

Morozov, Evgeny. 2011. *The Net Delusion: How Not to Liberate the World*. Penguin.

Morozov, Evgeny. 2013. *To Save Everything, Click Here: Technology, Solutionism, and the Urge to Fix Problems That Don't Exist*. Hardcover ed. Allen Lane.

Morris, Brian. 1994. *Anthropology of the Self: The Individual in Cultural Perspective*. Pluto Press.

Morse, Jack. 2019. "More than 1,000 Google Assistant Recordings Leaked, and Oh Boy." *Mashable*. Accessed July 22, 2019. https://mashable.com/article/google-assistant-recordings-leaked/.

Mosco, Vincent. 2004. *The Digital Sublime: Myth, Power and Cyberspace*. The MIT Press.

Moser, Carol, Tianying Chen, and Sarita Y. Schoenebeck. 2017. "Parents' And Children's Preferences about Parents Sharing About Children on Social Media." In *Proceedings of the 2017 CHI Conference on Human Factors in Computing Systems*, 5221–5225. ACM. https://doi.org/10.1145/3025453.3025587.

Mossberger, Karen, Caroline J. Tolbert, and Ramona S. McNeal. 2007. *Digital Citizenship: The Internet, Society, and Participation*. The MIT Press.

Nardi, Bonnie, and Vicki O'Day. 1999. "Information Ecologies: Using Technology with Heart: Chapter Four: Information Ecologies." *First Monday* 4 (5). http://firstmonday.org/ojs/index.php/fm/article/view/672.

Natale, Simone, and Andrea Ballatore. 2017. "Imagining the Thinking Machine: Technological Myths and the Rise of Artificial intelligence." *Convergence* 16, no.1: 3–18. https://doi.org/10.1177/1354856517715164.

New, Joshua. 2018. "How (and How Not) to Fix AI." *TechCrunch*, May 9, 2018. http://social.techcrunch.com/2018/07/26/how-and-how-not-to-fix-ai/.

Nissenbaum, Helen. 2010. *Privacy in Context: Technology, Policy, and the Integrity of Social Life*. Stanford University Press.

Nissenbaum, Helen. 2011. "A Contextual Approach to Privacy Online." *Daedalus*, Fall. https://www.amacad.org/publication/contextual-approach-privacy-online.

Noble, Safiya Umoja. 2018. *Algorithms of Oppression: How Search Engines Reinforce Racism*. NYU Press.

Nolas, Sevasti-Melissa, Christos Varvantakis, and Vinnarasan Aruldoss. 2016. "(Im) Possible Conversations? Activism, Childhood and Everyday Life." *Journal of Social and Political Psychology* 4, no. 1: 252–265.

O'Neil, Cathy. 2016. *Weapons of Math Destruction: How Big Data Increases Inequality and Threatens Democracy*. Crown/Archetype.

O'Reilly, Tim. 2005. *What Is Web 2.0: Design Patterns and Business Models for the Next Generation of Software*. O'Reilly Media.

Orwell, George. (1949) 2016. *1984*. Hamilton Books.

Oswell, David. 2013. *The Agency of Children: From Family to Global Human Rights*. Cambridge University Press.

Overdorf, Rebekah, Bogdan Kulynych, Ero Balsa, Carmela Troncoso, and Seda Gürses. 2018. "Questioning the Assumptions behind Fairness Solutions." *ArXiv* 1811.11293v1 [Cs]. http://arxiv.org/abs/1811.11293.

Ovia Health. 2018. "Ovia Privacy Policy." 2018. https://www.oviahealth.com/dynamic-privacy.

Ovuline Inc. 2019. "Ovia Pregnancy Tracker." App Store (promotional blurb). 2019. https://apps.apple.com/us/app/ovia-pregnancy-tracker/id719135369.

Ozga, Jenny. 2009. "Governing Education through Data in England: From Regulation to Self-evaluation." *Journal of Education Policy* 24, no. 2: 149–162. https://doi.org/10.1080/02680930902733121.

Pasquale, Frank. 2016. *The Black Box Society: The Secret Algorithms That Control Money and Information*. Reprint ed. Harvard University Press.

Pegg, David. 2018. "Labour Bought Data on 1m Mothers and Their Children." *The Guardian*, July 11, 2018. https://www.theguardian.com/politics/2018/jul/11/labour-bought-data-on-more-than-1m-mums-and-their-children-emmas-diary.

Perez, Francisco. 2016. "Story of Austrian Teen Suing Parents over Facebook Pictures Debunked | DW | 19.09.2016." *Deutsche Welle*, September 19, 2016. https://www.dw.com/en/story-of-austrian-teen-suing-parents-over-facebook-pictures-debunked/a-19562265.

Pfaffenberger, B. 1988. "Fetished Objects and Humanised Nature: Towards an Anthropology of Technology." *Man* 23, no. 2: 236–252. https://doi.org/10.2307/2802804.

Piersol, Kurt Wesley, and Gabriel Beddingfield. 2019. Pre-wakeword speech processing. US Patent 10,192,546, filed January 24, 2019. Accessed July 29, 2019. http://appft.uspto.gov/netacgi/nph-Parser?Sect1=PTO1&Sect2=HITOFF&d=PG01&p=1&u=%2Fnetahtml%2FPTO%2Fsrchnum.html&r=1&f=G&l=50&s1=%2220190156818%22.PGNR.&OS=DN/20190156818&RS=DN/20190156818.

Pink, Sarah, Debora Lanzeni, and Heather Horst. 2018. "Data Anxieties: Finding Trust in Everyday Digital Mess." *Big Data & Society* 5, no 1. https://doi.org/10.1177/2053951718756685.

Pink, Sarah, and Kerstin Leder Mackley. 2013. "Saturated and Situated: Expanding the Meaning of Media in the Routines of Everyday Life." *Media, Culture & Society* 35, no. 6: 677–691. https://doi.org/10.1177/0163443713491298.

Podesta, John. 2014. "Findings of the Big Data and Privacy Working Group Review." Whitehouse.gov. May 1, 2014. https://obamawhitehouse.archives.gov/blog/2014/05/01/findings-big-data-and-privacy-working-group-review.

Pratt, Jeff. 2003. *Class, Nation and Identity: The Anthropology of Political Movements*. Pluto Press.

Privacy International. 2018. "Privacy International Launches Campaign to Investigate Range of Data Companies That Facilitate Mass Data Exploitation." Press release, May 25, 2018. http://privacyinternational.org/press-release/2047/privacy-international-launches-campaign-investigate-range-data-companies.

Quan-Haase, Anabel, and Barry Wellman. 2006. "Hyperconnected Net Work: Computer-Mediated Community in a High-Tech Organization." In *The Firm as a Collaborative Community: Reconstructing Trust in the Knowledge Economy*, 281–333. Oxford University Press.

Quintin, Cooper. 2017. "The Pregnancy Panopticon." Electronic Frontier Foundation. https://www.eff.org/wp/pregnancy-panopticon.

Raghupathi, Wullianallur, and Viju Raghupathi. 2014. "Big Data Analytics in Healthcare: Promise and Potential." *Health Information Science and Systems* 2, no. 1: 3. https://doi.org/10.1186/2047-2501-2-3.

Rahman, Rajiur, and Chandan K. Reddy. 2015. "Electronic Health Records: A Survey." In *Healthcare Data Analytics*, edited by Chandan K. Reddy and Charu C. Aggarwal, 22–52. CRC Press.

Ramaswami, Prem. 2015. "A remedy for your health-related questions: Health info in the Knowledge Graph." *Official Google Blog* (blog). October 2, 2015. https://googleblog.blogspot.com/2015/02/health-info-knowledge-graph.html.

Redden, Joanna, and Jessica Brand. 2017. "Data Harm Record." Data Justice Lab, Cardiff University. https://datajusticelab.org/data-harm-record/.

Richardson, Rashida, Jason Schultz, and Kate Crawford. 2019. "Dirty Data, Bad Predictions: How Civil Rights Violations Impact Police Data, Predictive Policing Systems, and Justice." *New York University Law Review* 94: 192–233. Available at SSRN: https://papers.ssrn.com/abstract=3333423.

Rose, Megan. 2015. "The Average Parent Shares Almost 1,500 Images of Their Child Online before Their 5th Birthday." Parent Zone. https://parentzone.org.uk/article/average-parent-shares-almost-1500-images-their-child-online-their-5th-birthday.

Ross, Casey. 2019. "Amazon Alexa Now HIPAA-Compliant, Allows Secure Access to Data." *STAT Health Tech*, April 4, 2019. https://www.statnews.com/2019/04/04/amazon-alexa-hipaa-compliant/.

Ruppert, Evelyn, Engin Isin, and Didier Bigo. 2017. "Data Politics." *Big Data & Society* 4, no. 2: 2053951717717749. https://doi.org/10.1177/2053951717717749.

Russell, N. Cameron, Joel R. Reidenberg, Elizabeth Martin, and Thomas Norton. 2018. "Transparency and the Marketplace for Student Data." *Center on Law and Information Policy* 4 (June): 2–33.

Saltman, Kenneth J. 2016. *Scripted Bodies: Corporate Power, Smart Technologies, and the Undoing of Public Education*. Routledge.

Savirimuthu, Joseph. 2015. "Networked Children, Commercial Profiling and the EU Data Protection Reform Agenda: In the Child's Best Interests." In *The EU as a Children's Rights Actor: Law, Policy and Structural Dimensions*, edited by Ingi Iusmen and Helen Stalford, 221–257. Barbara Budrich.

Schechner, Sam, and Mark Secada. 2019. "You Give Apps Sensitive Personal Information. Then They Tell Facebook." *Wall Street Journal*, February 22, 2019. Tech. https://www.wsj.com/articles/you-give-apps-sensitive-personal-information-then-they-tell-facebook-11550851636.

Schofield Clark, Lynn. 2013. *The Parent App: Understanding Families in the Digital Age*. Oxford University Press.

Scism, Leslie. 2019. "New York Insurers Can Evaluate Your Social Media Use—If They Can Prove Why It's Needed." *Wall Street Journal*, January 30, 2019. https://www.wsj.com/articles/new-york-insurers-can-evaluate-your-social-media-useif-they-can-prove-why-its-needed-11548856802.

Scripps Research. 2018. "Transforming Pregnancy Research with a Smartphone App." *Scripps Research*. https://www.scripps.edu/news-and-events/press-room/2018/20180905-pregnancy-app-radin.html.

Segal, Howard P. 1985. *Technological Utopianism in American Culture*. Syracuse University Press.

Selwyn, Neil. 2013. *Distrusting Educational Technology: Critical Questions for Changing Times*. Routledge.

Singer, Natasha. 2018a. "For Sale: Survey Data on Millions of High School Students." *New York Times*, July 31, 2018. Business. https://www.nytimes.com/2018/07/29/business/for-sale-survey-data-on-millions-of-high-school-students.html.

Singer, Natasha. 2018b. "In Screening for Suicide Risk, Facebook Takes on Tricky Public Health Role." *New York Times*, December 30, 2018. Technology. https://www.nytimes.com/2018/12/31/technology/facebook-suicide-screening-algorithm.html.

Snap Inc. 2018. "Privacy Center." https://www.snap.com/en-US/privacy/privacy-policy.

Snap Inc. 2019. "Terms of Service." https://www.snap.com/en-US/terms/.

Solove, Daniel J. 2004. *The Digital Person: Technology and Privacy in the Information Age*. NYU Press.

Somé, Sobonfu. 1999. *Welcoming Spirit Home: Ancient African Teachings to Celebrate Children and Community*. New World Library.

Song, Indeok, Robert Larose, Matthew S. Eastin, and Carolyn A. Lin. 2004. "Internet Gratifications and Internet Addiction: On the Uses and Abuses of New Media." *CyberPsychology & Behavior* 7, no. 4: 384–394. https://doi.org/10.1089/cpb.2004.7.384.

Souto-Otero, Manuel, and Roser Beneito-Montagut. 2016. "From Governing through Data to Governmentality through Data: Artefacts, Strategies and the Digital Turn." *European Educational Research Journal* 15, no. 1: 14–33.

Spanu, Anca. 2018. "Google's Nest Is Quietly Making Its Way towards Digital Healthcare for the Elderly." *Health Care Weekly*. September 26, 2018. https://healthcareweekly.com/nest-googles-subsidiary-is-interested-in-digital-healthcare-for-the-elderly/.

Speed, Barbara. 2019. "Steps, Sips or Spending: Does the Trend for Tracking Our Habits Actually Make Our Lives Better?" *Prospect Magazine*, January 29, 2019. https://www.prospectmagazine.co.uk/magazine/barbara-speed-self-improvement-habits-fitbit-steps-alcohol-is-tracking-good-for-us.

Statt, Nick. 2018. "Amazon Told Employees It Would Continue to Sell Facial Recognition Software to Law Enforcement." *The Verge*, November 8, 2018. https://www.theverge.com/2018/11/8/18077292/amazon-rekognition-jeff-bezos-andrew-jassy-facial-recognition-ice-rights-violations.

Steinberg, Stacey B. 2017 "Sharenting: Children's Privacy in the Age of Social Media." *Emory Law Journal* 66: 839.

Stiegler, Bernard. 2009. "Teleologics of the Snail: The Errant Self Wired to a WiMax Network." *Theory, Culture & Society* 26, no. 2–3: 33–45. https://doi.org/10.1177/0263276409103105.

Stoilova, Mariya, Rishita Nandagiri, and Sonia Livingstone. 2019. "Children's Understanding of Personal Data and Privacy Online—a Systematic Evidence Mapping." *Information, Communication & Society*. Published Online First, September, 17, 2019. https://doi.org/10.1080/1369118X.2019.1657164.

Strangers, Yolande. 2016. "Envisioning the Smart Home: Reimagining a Smart Energy Future." In *Digital Materialities: Design and Anthropology*, edited by Sarah Pink, Elisenda Ardèvol, and Dèbora Lanzeni, 61–79. Bloomsbury Publishing.

Strauss, Valerie. 2018. "Students Protest Zuckerberg-Backed Digital Learning Program and Ask Him: 'What Gives You This Right?'" *Washington Post*, November 17, 2018. Answer Sheet. https://www.washingtonpost.com/education/2018/11/17/students-protest-zuckerberg-backed-digital-learning-program-ask-him-what-gives-you-this-right/.

Summit Learning. 2018. "Summit Learning Privacy Policy." https://cdn.summitlearning.org/marketing/privacy_center/privacy_policy.pdf.

Tate, Christie. 2019. "Perspective | My Daughter Asked Me to Stop Writing about Motherhood. Here's Why I Can't Do That." *Washington Post*, January 3, 2019. On Parenting: Perspective. https://www.washingtonpost.com/lifestyle/2019/01/03/my-daughter-asked-me-stop-writing-about-motherhood-heres-why-i-cant-do-that/.

Taylor, Linnet. 2017. "What Is Data Justice? The Case for Connecting Digital Rights and Freedoms Globally." *Big Data & Society* 4, no. 2: 2053951717736335. https://doi.org/10.1177/2053951717736335.

Terranova, Tiziana. 2004. *Network Culture: Politics for the Information Age*. Pluto Press.

Third, Amanda, and P. Collin. 2016. "Rethinking (Children's and Young People's) Citizenship through Dialogues on Digital Practice." In *Negotiating Digital Citizenship: Control, Contest and Culture*, edited by Anthony McCosker, Sonja Vivienne, and Amelia Johns, 41–59. Rowman & Littlefield International.

Thompson, E. P. 1967. "Time, Work-Discipline, and Industrial Capitalism." *Past & Present*, no. 38 (December): 56–97.

Thrift, Nigel. 1990. "The Making of a Capitalist Time Consciousness." In *The Sociology of Time*, edited by John Hassard. Palgrave Macmillan.

Timmerman, Peter. 2007. "Architecture in the Mirror of Technology." In *Tensions and Convergences: Technological And Aesthetic Transformations of Society*, edited by Reinhard Heil, Andreas Kaminski, Marcus Stippak, and Alexander Unger, 47–57. Transaction Publishers.

Tomlinson, John. 2007. *The Culture of Speed: The Coming of Immediacy*. Sage.

Travis, Alan. 2017. "May Pressured NHS to Release Data to Track Immigration Offenders." *The Guardian*, February 1, 2017. UK news. https://www.theguardian.com/uk-news/2017/feb/01/home-office-asked-former-nhs-digital-boss-to-share-data-to-trace-immigration-offenders.

Turkle, Sherry. 2011. *Alone Together: Why We Expect More from Technology and Less from Each Other*. Basic Books.

Turner, Victor. 1970. *The Forest of Symbols: Aspects of Ndembu Ritual*. Cornell University Press.

Turow, Joseph, Michael Hennessy, and Nora Draper. 2015. "The Tradeoff Fallacy: How Marketers Are Misrepresenting American Consumers and Opening Them Up to Exploitation." A Report from the Annenberg School of Communication, University of Pennsylvania. https://www.asc.upenn.edu/sites/default/files/TradeoffFallacy_1.pdf.

Twitter Inc. 2019a. "Twitter Privacy Policy." https://twitter.com/en/privacy.

Twitter Inc. 2019b. "Twitter Terms of Service." https://twitter.com/content/twitter-com/legal/en/tos.html.

UCAS (Universities and Colleges Admissions Services). 2019. "UCAS Media." UCAS Media UK. 2019. https://www.ucasmedia.com/node.

Usborne, Simon. 2018. "How the Hostile Environment Crept into UK Schools, Hospitals and Homes." *The Guardian*, August 1, 2018. UK news. https://www.theguardian.com/uk-news/2018/aug/01/hostile-environment-immigrants-crept-into-schools-hospitals-homes-border-guards.

US Department of Education. 2009. "Fact Sheet: Statewide Longitudinal Data Systems." Pamphlets. July 30, 2009. https://www2.ed.gov/programs/slds/factsheet.html.

Van Dijck, José, and David Nieborg. 2009. "Wikinomics and Its Discontents: A Critical Analysis of Web 2.0 Business Manifestos." *New Media & Society* 11, no. 5: 855–874. https://doi.org/10.1177/1461444809105356.

Vasel, Kathryn. 2019. "This Company Uses AI to Flag Racist and Sexist Comments from Potential Hires." *CNN*, Winter 2019. https://www.cnn.com/2019/04/12/success/fama-prescreen-employment/index.html.

Vertesi, Janet. 2014. "Internet Privacy and What Happens When You Try to Opt Out." *Time*, May 1. http://time.com/83200/privacy-internet-big-data-opt-out/.

Vlahos, James. 2015. "Barbie Wants to Get to Know Your Child." *New York Times*, September 16, 2015. Magazine. https://www.nytimes.com/2015/09/20/magazine/barbie-wants-to-get-to-know-your-child.html.

Wachowski, Lana, and Lilly Wachowski. 1999. *The Matrix*. Action, Sci-Fi. Warner Bros. Village Roadshow Pictures, Groucho Film Partnership.

Wainwright, David. 2006. "The Politics of Information and Communication Technology Diffusion: A Case Study in a UK Primary Health Care Trust." In *The Transfer and Diffusion of Information Technology for Organizational Resilience*, edited by Brian Donnellan, Tor Larsen, Linda Levine, and Janice DeGross, 71–89. Springer Science & Business Media.

Wang, Amy B. 2017. "Former Facebook VP Says Social Media Is Destroying Society with 'Dopamine-Driven Feedback Loops.'" *Washington Post*, December 12, 2017. https://www.washingtonpost.com/news/the-switch/wp/2017/12/12/former-facebook-vp-says-social-media-is-destroying-society-with-dopamine-driven-feedback-loops/.

Ward, Lucy. 2014. "Ucas Sells Access to Student Data for Phone and Drinks Firms' Marketing." *The Guardian*, March 12, 2014. Technology. https://www.theguardian.com/uk-news/2014/mar/12/ucas-sells-marketing-access-student-data-advertisers.

Weber, Max. 1978. *Economy and Society: An Outline of Interpretive Sociology*. University of California Press.

Weise, Elizabeth. 2018. "Amazon's Alexa Will Be Built into All New Homes from Lennar." *USA Today*. Accessed July 29, 2018. https://www.usatoday.com/story/tech/news/2018/05/09/amazons-alexa-built-into-all-new-homes-lennar/584004002/.

Whittaker, Freddie, and Billy Camden. 2016. "Non-White School Pupils Asked for Passports." *Schools Week*, September 23, 2016. https://schoolsweek.co.uk/pupils-who-were-not-white-british-told-to-send-in-birthplace-data/.

Williamson, Ben. 2017a. *Big Data in Education: The Digital Future of Learning, Policy and Practice*. Sage.

Williamson, Ben. 2017b. "Who Owns Educational Theory? Big Data, Algorithms and the Expert Power of Education Data Science." *E-Learning and Digital Media* 14, no. 3: 105–122. https://doi.org/10.1177/2042753017731238.

Wooldridge, Adrian. 2006. *Measuring the Mind: Education and Psychology in England c. 1860-1990*. Cambridge University Press.

Wyness, Michael. 2011. *Childhood and Society*. Macmillan International Higher Education.

Zarkadakis, George. 2015. *In Our Own Image: Will Artificial Intelligence Save or Destroy Us?* Rider.

Zuboff, Shoshana. 2015. "Big Other: Surveillance Capitalism and the Prospects of an Information Civilization." *Journal of Information Technology* 30, no. 1: 75–89. https://doi.org/10.1057/jit.2015.5.

Zuboff, Shoshana. 2019. *The Age of Surveillance Capitalism: The Fight for a Human Future at the New Frontier of Power*. Public Affairs.

Zuckerberg, Mark. 2016. "Building Jarvis." Facebook. 2016. https://www.facebook.com/notes/mark-zuckerberg/building-jarvis/10154361492931634/.

INDEX